影印版说明

本书是美国 McGraw-Hill Education 公司 2016 年出版的 *Handbook of Civil Engineering Calculations* Third Edition 的影印版，是目前最新的有关土木工程计算的指导手册，提供了与最新的规范和标准相一致的 3 000 多种计算方法，可以帮助读者准确地进行复杂的设计与施工计算。

考虑到使用方便，影印版分为 7 册：
1 结构工程
　Structural Engineering
2 钢筋混凝土与预应力混凝土工程及设计
　Reinforced and Prestressed Concrete Engineering and Design
3 木结构・土力学
　Timber Engineering・Soil Mechanics
4 测绘、路线设计及公路桥梁
　Surveying, Route Design and Highway Bridges
5 流体力学、水泵、管道及水电
　Fluid Mechanics, Pumps, Piping and Hydro Power
6 供水及雨水系统设计・污水处理及控制
　Water-Supply and Storm-Water System Design・Sanitary Wastewater Treatment and Control
7 工程经济
　Engineering Economics

作者泰勒・G・希克斯是一位经验丰富的顾问工程师，曾负责多个工厂的设计及运行，并在多所工程类院校任教和进行国际讲学，出版了 20 多部相关著作。

材料科学与工程图书工作室
联系电话　0451-86412421
　　　　　0451-86414559
邮　　箱　yh_bj@aliyun.com
　　　　　xuyaying81823@gmail.com
　　　　　zhxh6414559@aliyun.com

影印版

Third Edition

HANDBOOK OF CIVIL ENGINEERING CALCULATIONS
土木工程计算手册

结构工程
Structural Engineering

TYLER G. HICKS, P.E.

哈尔滨工业大学出版社
HARBIN INSTITUTE OF TECHNOLOGY PRESS

黑版贸审字08-2016-106号

Tyler G. Hicks
Handbook of Civil Engineering Calculations, Third Edition
ISBN 978-1-25-958685-9
Copyright © 2016 by McGraw-Hill Education.
All rights reserved. No part of this publication may be reproduced or transmitted in any form or by any means, electronic or mechanical, including without limitation photocopying, recording, taping, or any database, information or retrieval system, without the prior written permission of the publisher.
This authorized English reprint edition is jointly published by McGraw-Hill Education and Harbin Institute of Technology Press Co. Ltd.This edition is authorized for sale in the People's Republic of China only, excluding Hong Kong, Macao SAR and Taiwan.
Copyright © 2017 by McGraw-Hill Education.

版权所有。未经出版人事先书面许可，对本出版物的任何部分不得以任何方式或途径复制或传播，包括但不限于复印、录制、录音，或通过任何数据库、信息或可检索的系统。

本授权英文影印版由麦格劳—希尔教育出版公司和哈尔滨工业大学出版社有限公司合作出版。此版本经授权仅限在中华人民共和国境内（不包括香港特别行政区、澳门特别行政区和台湾地区）销售。

版权©2017由麦格劳—希尔教育出版公司所有。

本书封面贴有McGraw-Hill Education公司防伪标签，无标签者不得销售。

图书在版编目（CIP）数据

土木工程计算手册.结构工程=Handbook of Civil Engineering Calculations. Structural Engineering：英文/（美）泰勒·G·希克斯（Tyler G. Hicks）主编.—影印本.—哈尔滨：哈尔滨工业大学出版社，2017.3
ISBN 978-7-5603-6342-4

Ⅰ.①土… Ⅱ.①泰… Ⅲ.①土木工程－工程计算－技术手册－英文②土木结构－结构工程－工程计算－技术手册－英文 Ⅳ.①TU-32

中国版本图书馆CIP数据核字（2017）第041744号

责任编辑	杨 桦　许雅莹　张秀华
出版发行	哈尔滨工业大学出版社
社　　址	哈尔滨市南岗区复华四道街10号 邮编 150006
传　　真	0451-86414749
网　　址	http://hitpress.hit.edu.cn
印　　刷	哈尔滨市石桥印务有限公司
开　　本	660mm×980mm 1/16 印张 13.25
版　　次	2017年3月第1版 2017年3月第1次印刷
书　　号	ISBN 978-7-5603-6342-4
定　　价	68.00元

（如因印刷质量问题影响阅读，我社负责调换）

HANDBOOK OF CIVIL ENGINEERING CALCULATIONS

Tyler G. Hicks, P.E., Editor
International Engineering Associates
Member: American Society of Mechanical Engineers
United States Naval Institute

S. David Hicks, Coordinating Editor

George K. Korley, P.E.
President/CEO
Korley Engineering Consultants LLC
New York, N.Y.
Contributor, LRFD, Section 1

Third Edition

New York Chicago San Francisco Athens London
Madrid Mexico City Milan New Delhi
Singapore Sydney Toronto

CONTENTS

Preface vii
How to Use This Handbook xi

第 1 册（本册）

Section 1. Structural Engineering 1.1

第 2 册

Section 2. Reinforced and Prestressed Concrete Engineering and Design 2.1

第 3 册

Section 3. Timber Engineering 3.1

Section 4. Soil Mechanics 4.1

第 4 册

Section 5. Surveying, Route Design, and Highway Bridges 5.1

第 5 册

Section 6. Fluid Mechanics, Pumps, Piping, and Hydro Power 6.1

第 6 册

Section 7. Water-Supply and Storm-Water System Design 7.1

Section 8. Sanitary Wastewater Treatment and Control 8.1

第 7 册

Section 9. Engineering Economics 9.1

Index I.1

PREFACE

This third edition of this handbook has been thoroughly updated to reflect the new developments in civil engineering since the publication of the second edition.

Thus, Section 1 contains 35 new LRFD (load resistance factor design) calculation procedures. The new procedures show the civil engineer how to use LRFD in his or her daily design work, and since LRFD is a preferred design method among recent civil engineering graduates, the handbook is right in line with current design methods.

However, since an alternative design method is still used by some engineers around the world, namely ASD—Allowable Stress Design—it, too, is covered in this handbook. Thus the civil engineer will find both methods available to him or her so a suitable design method can be chosen using the third edition of this handbook.

Section 2, on reinforced and prestressed concrete, has been updated with new calculation procedures. The calculation procedures given in this handbook provide a wide-ranging coverage of this important specialty in civil engineering.

Section 3, timber engineering, has new methods and procedures for making calculations in this popular branch of civil engineering.

Section 4, soil mechanics, provides new ways to make soil calculations. Since nearly all structures require soil calculations, the new method given is an important addition to this handbook.

Section 5, surveying, route design, and highway bridges, has important new calculation procedures. Highways and roads, in much of the civilized world, are in poor condition. Potholes, cracks, and edge deterioration are rampant. Drivers often suffer blown tires, broken axles, and damaged wheels on such highways and roads.

When complaints are filed, the most frequent answer heard is "Roadway budgets have been cut; we don't have the money for repairs." In the United States, Congress is considering a major highway repair bill, but it seems to be moving very slowly—much slower than it should—considering the poor conditions of highways and local roads.

If the bill passes and money is appropriated for highway and road repairs, an enormous amount of work will be required. Civil engineers will have at least 10 years of highway and road work ahead of them. And, as part of the bill, new highways will be built. This will open opportunities for creative designs that make for safer driving in all types of weather. Truly, an exciting future is open to civil engineers in highway and road work.

As part of highway and road repairs, bridges will also need lots of work, plus—in some cases—complete replacement. Bridges have suffered numerous failures in recent years, with unfortunate loss of life. Many highway bridges need extensive repairs to steel elements, along with reinforced concrete rehabilitation.

Again, money is needed for this important work. And, like highways, the excuse is lack of funding. Hopefully, money assigned to highway repairs will include funds for bridge repair and replacement.

With bridge design constantly improving and new repair methods being developed, highway spans will be upgraded to reduce the loss of life from structural failures. The third edition of this handbook includes much valuable material on safe and effective bridge design. Civil engineers are urged to become familiar with codes covering safe bridge design. Knowing, and using, such codes will enhance almost every design, while making it safer for drivers and pedestrians.

With the highway, road, and bridge work ahead in next one or two decades, civil engineers have a very bright future. These engineers will seldom want for an important position. The demand for their skills will be ongoing for a very long time.

Section 6, fluid mechanics, pumps, piping, and hydro power, has new calculation procedures in it. A new method of generating hydro power in drinking water, irrigation, and wastewater lines is presented in this section. Also, a proven way for preliminary hydro power generating unit selection is given.

Section 7, water supply and storm-water system design, presents new ideas on water supply and usage. Water shortages are plaguing many parts of the world today. And these water shortages go far beyond the drinking water supply. In some countries the water shortage extends to farms. Owners of farms do not have enough natural water to irrigate their crops. So, some farmers move from nearly barren lands to more copiously supplied areas where they can safely grow their crops.

Drinking water supplies, covered in this section of this handbook, are also in short supply in some countries. Such shortages are critical to people's health. Finding sufficient potable water and delivering it to people's homes and work places is a critical task for civil engineers. The design of safe drinking water systems is also presented in this section. Thus, the civil engineer has the tools to provide safe drinking water, regardless of its source.

And since storm waters may be a source of drinking water, and a scourge to roads and buildings, its control is also provided. Full coverage of storm-water runoff rates, rainfall intensity, and the sizing of sewer pipes is also given in this section. Thus, the civil engineer is fully prepared to both supply and control water in all its uses and sources.

Section 8 covers sanitary wastewater treatment and control—an important topic in today's environmentally conscious world. The newest treatment methods are discussed in a number of important calculation procedures.

Section 9 covers engineering economics. This section has been fully revised so it focuses on the key calculations the civil engineer must make during his or her career. The result is a laser-like focus on the important calculation procedures used in daily civil engineering practice.

Since the second edition of this book was published, there have been major changes in civil engineering. These changes include following:

- **Antiterrorism construction** to protect large and small buildings structurally against terror attacks such as those that have occurred in New York, Paris, Mali, etc.
- **Increased building security features** are now included in almost every major—and many minor—structures to which the public has access. The increased security is to prevent internal and external sabotage and terrorism that could endanger the occupants and the structure.
- **Building Codes have been changed** to provide better protection for occupants and structures. These Code changes affect the daily design procedures of many civil engineers. Such changes are reflected in a number of the calculation procedures in this handbook.
- **"Green" building design is more popular than ever.** Buildings now win awards for their "green" efficiency that reduces energy consumption in new, existing, and rehabilitated buildings. Such "green" awards are important leasing sales features.
- **Major steps to improve indoor air quality (IAQ)** for all buildings have been taken in building design. These steps include much more than prohibiting occupants from smoking in buildings. New rules prohibit smoking within stated distances of the exterior of buildings. IAQ is a major concern in office buildings, schools, hotels, motels, factories, and other buildings throughout the world.
- **Better hurricane, tornado, flood, and wave resistance design of buildings** and other public structures after several disasters including Hurricanes Katrina and Sandy, the

superstorms in the Pacific and Indian Ocean. The loss of more than 250,000 lives in such storms has civil engineers searching for better ways to design, and build, structure—buildings, bridges, dams, etc.—to withstand the enormous forces of nature while protecting occupants. Also under study are (a) early-warning systems to alert people to the onset of dangerous conditions and (b) better escape routes for people fleeing affected areas. New approaches to levee and flood wall design, especially in cities like New Orleans, are being used to prevent recurrence of Hurricane Katrina losses. All these changes will be the work of civil engineers, with the assistance of other specialized professionals.

Taller buildings are being constructed in major cities around the worid. Thus, 1,000-foot+ (305 m) mixed-use buildings (residences, stores, offices) are being constructed worldwide. Some even have wind turbines to generate the electric power needed to run the building. Civil engineers will be busy designing the foundations, structural members, and reinforced and prestressed concrete elements of these extra-tall buildings for years to come.

And with the emphasis on "clean energy," hydro power sites are being developed at a faster pace than in many previous decades. New, and updated, generating facilities offer more than just electricity. Some serve as water-supply sources for fresh water used in both agricultural and domestic systems. Sites being developed are in areas throughout the world because electricity is needed almost everywhere on the globe. Civil engineers have, and will continue to have, a major role in the design and construction of these new, and updated, hydro power facilities.

With so many changes "on the drawing board" and computer screen, civil engineers and designers are seeking ways to include the advancements in their current and future design of buildings, bridges, dams, and other structures. This third edition includes many of the proposed changes so designers can include them in their thinking and calculations.

While there are computer programs that help the civil engineer with a variety of engineering calculations, such programs are highly specialized and do not have the breadth of coverage this handbook provides. Further, such computer programs are usually expensive. Because of their high cost, these computer programs can be justified only when a civil engineer makes a number of repetitive calculations on almost a daily basis. In contrast, this handbook can be used in the office, field, drafting room, or laboratory. It provides industry-wide coverage in a convenient and affordable package. As such, this handbook fills a long-existing need felt by civil engineers worldwide.

In contrast, civil engineers using civil-engineering computer programs often find data-entry time requirements excessive for quick one-off-type calculations. When one-off-type calculations are needed, most civil engineers today turn to their electronic calculator, desktop, or laptop computer and perform the necessary steps to obtain the solution desired. But where repetitive calculations are required, a purchased computer program will save time and energy in the usual medium-size or large civil-engineering design office. Small civil-engineering offices generally resort to manual calculation for even repetitive procedures because the investment for one or more major calculation programs is difficult to justify in economic terms.

Even when purchased computer programs are extensively used, careful civil engineers still insist on manually checking results on a random basis to be certain the program is accurate. This checking can be speeded by any of the calculation procedures given in this handbook. Many civil engineers remark to the author that they feel safer, knowing they have manually verified the computer results on a spot-check basis. With liability for civil engineering designs extending beyond the lifetime of the designer, every civil engineer seeks the "security blanket" provided by manual verification of the results furnished by a computer program run on a desktop, laptop, or workstation computer. This handbook gives the tools needed for manual verification of some 2000 civil engineering calculation procedures.

Each section in this handbook is written by one or more experienced professional engineers who is a specialist in the field covered. The contributors draw on their wide experience in their field to give each calculation procedure an in-depth coverage of its topic. So the person using the procedure gets step-by-step instructions for making the calculation plus background information on the subject that is the topic of the procedure.

And because the handbook is designed for worldwide use, both earlier and more modern topics are covered. For example, the handbook includes concise coverage of riveted girders, columns, and connections. While today's civil engineer may say that riveted construction is a method long past its prime, there are millions of existing structures worldwide that were built using rivets. So when a civil engineer is called on to expand, rehabilitate, or tear down such a structure, he or she must be able to analyze the riveted portions of the structure. This handbook provides that capability in a convenient and concise form.

In the realm of modern design techniques, the load and resistance factor method (LRFD) is covered with more than 30 calculation procedures showing its use in various design situations. The LRFD method is ultimately expected to replace the well-known and widely used allowable stress design (ASD) method for structural steel building frameworks. In today's design world, many civil engineers are learning the advantages of the LRFD method and growing to prefer it over the ASD method.

Also included in this handbook is a comprehensive section titled "How to Use This Handbook." It details the variety of ways a civil engineer can use this handbook in his or her daily engineering work. Included as part of this section are steps showing the civil engineer how to construct a private list of SI conversion factors for the specific work the engineer specializes in.

The step-by-step *practical* and *applied* calculation procedures in this handbook are arranged so they can be followed by anyone with an engineering or scientific background. Each worked-out procedure presents fully explained and illustrated steps for solving similar problems in civil engineering design, research, field, academic, or license-examination situations. For any applied problem, all the civil engineer needs to do is place his or her calculation sheets alongside this handbook and follow the step-by-step procedure line for line to obtain the desired solution for the actual real-life problem. By following the calculation procedures in this handbook, the civil engineer, scientist, or technician will obtain accurate results in minimum time with least effort. And the approaches and solutions presented are modern throughout.

The editor hopes this handbook is helpful to civil engineers worldwide. If the handbook user finds procedures that belong in the book but have been left out, the editor urges the engineer to send the title of the procedure to him, in care of the publisher. If the procedure is useful, the editor will ask for the entire text. And if the text is publishable, the editor will include the calculation procedure in the next edition of the handbook. Full credit will be given to the person sending the procedure to the editor. And if users find any errors in the handbook, the editor will be grateful for having these called to his attention. Such errors will be corrected in the next printing of the handbook. In closing, the editor hopes that civil engineers worldwide find this handbook helpful in their daily work.

TYLER G. HICKS, P.E.

HOW TO USE THIS HANDBOOK

There are two ways to enter this handbook to obtain the maximum benefit from the time invested. The first entry is through the index; the second is through the table of contents of the section covering the discipline, or related discipline, concerned. Each method is discussed in detail below.

Index. Great care and considerable time were expended on preparation of the index of this handbook so that it would be of maximum use to every reader. As a general guide, enter the index using the generic term for the type of calculation procedure being considered. Thus, for the design of a beam, enter at *beam(s)*. From here, progress to the specific type of beam being considered—such as *continuous, of steel*. Once the page number or numbers of the appropriate calculation procedure are determined, turn to them to find the step-by-step instructions and worked-out example that can be followed to solve the problem quickly and accurately.

Contents. The contents at the beginning of each section lists the titles of the calculation procedures contained in that section. Where extensive use of any section is contemplated, the editor suggests that the reader might benefit from an occasional glance at the table of contents of that section. Such a glance will give the user of this handbook an understanding of the breadth and coverage of a given section, or a series of sections. Then, when he or she turns to this handbook for assistance, the reader will be able more rapidly to find the calculation procedure he or she seeks.

Calculation Procedures. Each calculation procedure is a unit in itself. However, any given calculation procedure will contain subprocedures that might be useful to the reader. Thus, a calculation procedure on pump selection will contain subprocedures on pipe friction loss, pump static and dynamic heads, etc. Should the reader of this handbook wish to make a computation using any of such subprocedures, he or she will find the worked-out steps that are presented both useful and precise. Hence, the handbook contains numerous valuable procedures that are useful in solving a variety of applied civil engineering problems.

One other important point that should be noted about the calculation procedures presented in this handbook is that many of the calculation procedures are equally applicable in a variety of disciplines. Thus, a beam-selection procedure can be used for civil-, chemical-, mechanical-, electrical-, and nuclear-engineering activities, as well as some others. Hence, the reader might consider a temporary neutrality for his or her particular specialty when using the handbook because the calculation procedures are designed for universal use.

Any of the calculation procedures presented can be programmed on a computer. Such programming permits rapid solution of a variety of design problems. With the growing use of low-cost time sharing, more engineering design problems are being solved using a remote terminal in the engineering office. The editor hopes that engineers throughout the world will make greater use of work stations and portable computers in solving applied engineering problems. This modern equipment promises greater speed and accuracy for nearly all the complex design problems that must be solved in today's world of engineering.

To make the calculation procedures more amenable to computer solution (while maintaining ease of solution with a handheld calculator), a number of the algorithms in the handbook have been revised to permit faster programming in a computer environment. This enhances ease of solution for any method used—work station, portable computer, or calculator.

SI Usage. The technical and scientific community throughout the world accepts the SI (System International) for use in both applied and theoretical calculations. With such widespread acceptance of SI, every engineer must become proficient in the use of this system of units if he or she is to remain up-to-date. For this reason, every calculation procedure in this handbook is given in both the United States Customary System (USCS) and SI. This will help all engineers become proficient in using both systems of units. In this handbook the USCS unit is generally given first, followed by the SI value in parentheses or brackets. Thus, if the USCS unit is 10 ft, it will be expressed as 10 ft (3 m).

Engineers accustomed to working in USCS are often timid about using SI. There really aren't any sound reasons for these fears. SI is a logical, easily understood, and readily manipulated group of units. Most engineers grow to prefer SI, once they become familiar with it and overcome their fears. This handbook should do much to "convert" USCS-user engineers to SI because it presents all calculation procedures in both the known and unknown units.

Overseas engineers who must work in USCS because they have a job requiring its usage will find the dual-unit presentation of calculation procedures most helpful. Knowing SI, they can easily convert to USCS because all procedures, tables, and illustrations are presented in dual units.

Learning SI. An efficient way for the USCS-conversant engineer to learn SI follows these steps:

1. List the units of measurement commonly used in your daily work.
2. Insert, opposite each USCS unit, the usual SI unit used; Table 1 shows a variety of commonly used quantities and the corresponding SI units.
3. Find, from a table of conversion factors, such as Table 2, the value to use to convert the USCS unit to SI, and insert it in your list. (Most engineers prefer a conversion factor that can be used as a multiplier of the USCS unit to give the SI unit.)
4. Apply the conversion factors whenever you have an opportunity. Think in terms of SI when you encounter a USCS unit.
5. Recognize—here and now—that the most difficult aspect of SI is becoming comfortable with the names and magnitude of the units. Numerical conversion is simple, once you've set up *your own* conversion table. So think Pascal whenever you encounter pounds per square inch pressure, Newton whenever you deal with a force in pounds, etc.

SI Table for a Civil Engineer. Let's say you're a civil engineer and you wish to construct a conversion table and SI literacy document for yourself. List the units you commonly meet in your daily work; Table 1 is the list compiled by one civil engineer. Next, list the SI unit equivalent for the USCS unit. Obtain the equivalent from Table 2. Then, using Table 2 again, insert the conversion multiplier in Table 1.

Keep Table 1 handy at your desk and add new units to it as you encounter them in your work. Over a period of time you will build a personal conversion table that will be valuable to you whenever you must use SI units. Further, since *you* compiled the table, it will have a familiar and nonfrightening look, which will give you greater confidence in using SI.

TABLE 1. Commonly Used USCS and SI Units*

USCS unit	SI unit	SI symbol	Conversion factor—multiply USCS unit by this factor to obtain the SI unit
square feet	square meters	m^2	0.0929
cubic feet	cubic meters	m^3	0.2831
pounds per square inch	kilopascal	kPa	6.894
pound force	newton	N	4.448
foot pound torque	newton-meter	Nm	1.356
kip-feet	kilo-newton	kNm	1.355
gallons per minute	liters per second	L/s	0.06309
kips per square inch	megapascal	MPa	6.89
inch	millimeter	mm	25.4
feet	millimeter	mm	304.8
	meter	m	0.3048
square inch	square millimeter	mm^2	0.0006452
cubic inch	cubic millimeter	mm^3	0.00001638
$inch^4$	$millimeter^4$	mm^4	0.000000416
pound per cubic foot	kilogram per cubic meter	kg/m^3	16.0
pound per foot	kilogram per meter	kg/m	1.49
pound per foot force	Newton per meter	N/m	14.59
pound per inch force	Newton per meter	N/m	175.1
pound per foot density	kilogram per meter	kg/m	1.488
pound per inch density	kilogram per meter	kg/m	17.86
pound per square inch load concentration	kilogram per square meter	kg/m^2	703.0
pound per square foot load concentration	kilogram per square meter	kg/m^2	4.88
pound per square foot pressure	Pascal	Pa	47.88
inch-pound torque	Newton-meter	N-m	0.1129
chain	meter	m	20.117
fathom	meter	m	1.8288
cubic foot per second	cubic meter per second	m^3/s	0.02831
$foot^4$ (area moment of inertia)	$meter^4$	m^4	0.0086309
mile	meter	m	0.0000254
square mile	square meter	m^2	2589998.0
pound per gallon (UK liquid)	kilogram per cubic meter	kg/m^3	99.77
pound per gallon (U.S. liquid)	kilogram per cubic meter	kg/m^3	119.83
poundal	Newton	N	0.11382
square (100 square feet)	square meter	m^2	9.29
ton (long 2,240 lb)	kilogram	kg	1016.04
ton (short 2,000 lb)	kilogram	kg	907.18
ton, short, per cubic yard	kilogram per cubic meter	kg/m^3	1186.55
ton, long, per cubic yard	kilogram per cubic meter	kg/m^3	1328.93
ton force (2,000 lbf)	Newton	N	8896.44
yard, length	meter	m	0.0914

(continued)

TABLE 1. Commonly Used USCS and SI Units* (*Continued*)

USCS unit	SI unit	SI symbol	Conversion factor—multiply USCS unit by this factor to obtain the SI unit
square yard	square meter	m^2	0.08361
cubic yard	cubic meter	m^3	0.076455
acre feet	cubic meter	m^3	1233.49
acre	square meter	m^2	4046.87
cubic foot per minute	cubic meter per second	m^3/s	0.0004719

*Because of space limitations this table is abbreviated. For a typical engineering practice an actual table would be many times this length.

TABLE 2. Typical Conversion Table*

To convert from	To	Multiply by	
square feet	square meters	9.290304	E - 02
foot per second squared	meter per second squared	3.048	E - 01
cubic feet	cubic meters	2.831685	E - 02
pound per cubic inch	kilogram per cubic meter	2.767990	E + 04
gallon per minute	liters per second	6.309	E - 02
pound per square inch	kilopascal	6.894757	
pound force	Newton	4.448222	
kip per square foot	Pascal	4.788026	E + 04
acre-foot per day	cubic meter per second	1.427641	E - 02
acre	square meter	4.046873	E + 03
cubic foot per second	cubic meter per second	2.831685	E - 02

Note: The E indicates an exponent, as in scientific notation, followed by a positive or negative number, representing the power of 10 by which the given conversion factor is to be multiplied before use. Thus, for the square feet conversion factor, 9.290304 × 1/100 = 0.09290304, the factor to be used to convert square feet to square meters. For a positive exponent, as in converting acres to square meters, multiply by 4.046873 × 1000 = 4046.8.

Where a conversion factor cannot be found, simply use the dimensional substitution. Thus, to convert pounds per cubic inch to kilograms per cubic meter, find 1 lb = 0.4535924 kg, and 1 in^3 = 0.00001638706 m^3. Then, 1 lb/in^3 = 0.4535924 kg/0.00001638706 m^3 27,680.01, or 2.768 × E + 4.

*This table contains only selected values. See the U.S. Department of the Interior *Metric Manual*, or National Bureau of Standards, *The International System of Units* (SI), both available from the U.S. Government Printing Office (GPO), for far more comprehensive listings of conversion factors.

Units Used. In preparing the calculation procedures in this handbook, the editors and contributors used standard SI units throughout. In a few cases, however, certain units are still in a state of development. For example, the unit *tonne* is used in certain industries, such as waste treatment. This unit is therefore used in the waste treatment section of this handbook because it represents current practice. However, only a few SI units are still under development. Hence, users of this handbook face little difficulty from this situation.

Computer-aided Calculations. Widespread availability of programmable pocket calculators and low-cost laptop computers allows engineers and designers to save thousands of hours of calculation time. Yet each calculation procedure must be programmed, unless the engineer is willing to use off-the-shelf software. The editor—observing thousands of engineers over the years—detects reluctance among technical personnel to use untested and unproven software programs in their daily calculations. Hence, the tested and proven procedures in this handbook form excellent programming input for programmable pocket calculators, laptop computers, minicomputers, and mainframes.

A variety of software application programs can be used to put the procedures in this handbook on a computer. Typical of these are MathSoft, Algor, and similar programs.

There are a number of advantages for the engineer who programs his or her own calculation procedures, namely: (1) The engineer knows, understands, and approves *every* step in the procedure; (2) there are *no* questionable, unknown, or legally worrisome steps in the procedure; (3) the engineer has complete faith in the result because he or she knows every component of it; and (4) if a variation of the procedure is desired, it is relatively easy for the engineer to make the needed changes in the program, using this handbook as the source of the steps and equations to apply.

Modern computer equipment provides greater speed and accuracy for almost all complex design calculations. The editor hopes that engineers throughout the world will make greater use of available computing equipment in solving applied engineering problems. Becoming computer literate is a necessity for every engineer, no matter which field he or she chooses as a specialty. The procedures in this handbook simplify every engineer's task of becoming computer literate because the steps given comprise—to a great extent—the steps in the computer program that can be written.

SECTION 1
STRUCTURAL ENGINEERING

GEORGE K. KORLEY, P.E.
President/CEO
Korley Engineering Consultants LLC

MAX KURTZ, P.E.
Consulting Engineer

Load and Resistance Factor Design Overview	1.4
Part 1: Structural Steel Design	
STEEL BEAMS AND PLATE GIRDERS	1.8
Most Economic Section for a Beam with a Continuous Lateral Support under a Uniform Load	1.9
Most Economic Section for a Beam with an Intermittent Lateral Support under a Uniform Load	1.10
Design of a Beam with Reduced Moment Resistance	1.11
Design of a Cover-Plated Beam	1.13
Design of a Continuous Beam	1.17
Shearing Stress in a Beam—Exact Method	1.18
Shearing Stress in a Beam—Approximate Method	1.19
Moment Capacity of a Welded Plate Girder	1.19
Analysis of a Riveted Plate Girder	1.21
Design of a Welded Plate Girder	1.22
STEEL COLUMNS AND TENSION MEMBERS	1.26
Capacity of a Built-Up Column	1.27
Capacity of a Double-Angle Star Strut	1.28
Section Selection for a Column with Two Effective Lengths	1.29
Stress in Column with Partial Restraint against Rotation	1.30
Lacing of Built-Up Column	1.31
Selection of a Column with a Load at an Intermediate Level	1.32
Design of an Axial Member for Fatigue	1.33
Investigation of a Beam Column	1.34
Application of Beam-Column Factors Using ASD Method	1.35
Net Section of a Tension Member	1.36
Design of a Double-Angle Tension Member	1.36
LOAD AND RESISTANCE FACTOR METHOD	1.38
Determining if a Given Beam is Compact or Noncompact	1.39
Determining Column Axial Shortening with a Specified Load	1.40
Determining the Compressive Strength of a Welded Section	1.41
Determining Beam Flexural Design Strength for Minor- and Major-Axis Bending	1.42
Designing Web Stiffeners for Welded Beams	1.43

Determining the Design Moment and Shear Strength of a Built-Up Wide-Flange Welded Beam Section	1.44
Finding the Lightest Section to Support a Specified Load	1.47
Combined Flexure and Compression in Beam-Columns in a Braced Frame	1.49
Simplified Second-Order Analysis	1.51
Selection of Concrete-Filled Steel Column	1.55
Determining Design Compressive Strength of Composite Columns	1.56
Analyzing a Concrete Slab for Composite Action	1.57
Determining the Design Shear Strength of a Beam Web	1.58
Designing a Bearing Plate for a Beam and It's End Reaction	1.59
Determining Beam Length to Eliminate Bearing Plate	1.61
ANALYSIS OF STRESS AND STRAIN	**1.62**
Stress Caused by an Axial Load	1.62
Deformation Caused by an Axial Load	1.62
Deformation of a Built-Up Member	1.62
Reactions at Elastic Supports	1.63
Analysis of Cable Supporting a Concentrated Load	1.65
Displacement of Truss Joint	1.65
Axial Stress Caused by Impact Load	1.66
Stresses on an Oblique Plane	1.67
Evaluation of Principal Stresses	1.68
Hoop Stress in Thin-Walled Cylinder under Pressure	1.69
Stresses in Prestressed Cylinder	1.70
Hoop Stress in Thick-Walled Cylinder	1.70
Thermal Stress Resulting from Heating a Member	1.71
Thermal Effects in Composite Member Having Elements in Parallel	1.72
Thermal Effects in Composite Member Having Elements in Series	1.73
Shrink-Fit Stress and Radial Pressure	1.73
Torsion of a Cylindrical Shaft	1.74
Analysis of a Compound Shaft	1.74
STRESSES IN FLEXURAL MEMBERS	**1.75**
Shear and Bending Moment in a Beam	1.76
Beam Bending Stresses	1.77
Analysis of a Beam on Movable Supports	1.78
Flexural Capacity of a Compound Beam	1.79
Analysis of a Composite Beam	1.80
Beam Shear Flow and Shearing Stress	1.82
Locating the Shear Center of a Section	1.83
Bending of a Circular Flat Plate	1.84
Bending of a Rectangular Flat Plate	1.85
Combined Bending and Axial Load Analysis	1.85
Flexural Stress in a Curved Member	1.87
Soil Pressure under Dam	1.87
Load Distribution in Pile Group	1.88
DEFLECTION OF BEAMS	**1.89**
Double-Integration Method of Determining Beam Deflection	1.89
Moment-Area Method of Determining Beam Deflection	1.90
Conjugate-Beam Method of Determining Beam Deflection	1.91
Unit-Load Method of Computing Beam Deflection	1.92
Deflection of a Cantilever Frame	1.93
STATICALLY INDETERMINATE STRUCTURES	**1.95**
Shear and Bending Moment of a Beam on a Yielding Support	1.95
Maximum Bending Stress in Beams Jointly Supporting a Load	1.96
Theorem of Three Moments	1.97
Theorem of Three Moments: Beam with Overhang and Fixed End	1.98
Bending-Moment Determination by Moment Distribution	1.99
Analysis of a Statically Indeterminate Truss	1.101

MOVING LOADS AND INFLUENCE LINES	1.103
Analysis of Beam Carrying Moving Concentrated Loads	1.103
Influence Line for Shear in a Bridge Truss	1.104
Force in Truss Diagonal Caused by a Moving Uniform Load	1.106
Force in Truss Diagonal Caused by Moving Concentrated Loads	1.106
Influence Line for Bending Moment in Bridge Truss	1.108
Force in Truss Chord Caused by Moving Concentrated Loads	1.109
Influence Line for Bending Moment in Three-Hinged Arch	1.110
Deflection of a Beam under Moving Loads	1.112
RIVETED AND WELDED CONNECTIONS	1.112
Capacity of a Rivet	1.113
Investigation of a Lap Splice	1.114
Design of a Butt Splice	1.115
Design of a Pipe Joint	1.116
Moment on Riveted Connection	1.117
Eccentric Load on Riveted Connection	1.118
Design of a Welded Lap Joint	1.120
Eccentric Load on a Welded Connection	1.121
PLASTIC DESIGN OF STEEL STRUCTURES	1.122
Allowable Load on Bar Supported by Rods	1.122
Determination of Section Shape Factors	1.123
Determination of Ultimate Load by the Static Method	1.124
Determining the Ultimate Load by the Mechanism Method	1.126
Analysis of a Fixed-End Beam under Concentrated Load	1.127
Analysis of a Two-Span Beam with Concentrated Loads	1.128
Selection of Sizes for a Continuous Beam	1.129
Mechanism-Method Analysis of a Rectangular Portal Frame	1.131
Analysis of a Rectangular Portal Frame by the Static Method	1.134
Theorem of Composite Mechanisms	1.134
Analysis of an Unsymmetric Rectangular Portal Frame	1.135
Analysis of Gable Frame by Static Method	1.137
Theorem of Virtual Displacements	1.139
Gable-Frame Analysis by Using the Mechanism Method	1.140
Reduction in Plastic-Moment Capacity Caused by Axial Force	1.141
Part 2: Hangers, Connectors, and Wind-Stress Analysis	
Design of an Eyebar	1.144
Analysis of a Steel Hanger	1.145
Analysis of a Gusset Plate	1.146
Design of a Semirigid Connection	1.147
Riveted Moment Connection	1.149
Design of a Welded Flexible Beam Connection	1.151
Design of a Welded Seated Beam Connection	1.153
Design of a Welded Moment Connection	1.154
Rectangular Knee of Rigid Bent	1.155
Curved Knee of Rigid Bent	1.157
Base Plate for Steel Column Carrying Axial Load	1.158
Base for Steel Column with End Moment	1.158
Grillage Support for Column	1.160
Wind-Stress Analysis by Portal Method	1.162
Wind-Stress Analysis by Cantilever Method	1.164
Wind-Stress Analysis by Slope-Deflection Method	1.167
Wind Drift of a Building	1.169
Reduction in Wind Drift by Using Diagonal Bracing	1.171
Light-Gage Steel Beam with Unstiffened Flange	1.172
Light-Gage Steel Beam with Stiffened Compression Flange	1.173

LOAD AND RESISTANCE FACTOR DESIGN OVERVIEW

Load and resistance factor design (LRFD) incorporates state-of-the-art analysis and design methodologies with load and resistance factors based on the known *variability* of applied loads and material properties. These load and resistance factors are *calibrated* from actual statistics to ensure a uniform level of safety.

LRFD provides reliability-based limit state specifications, uses probabilistic methods to derive loads and resistance factors and provides uniform reliability in design and load ratings.

LRFD Limit States

A *limit state* is a *defined condition* beyond which a structural component ceases to satisfy the provisions for which it is designed.

Resistance is a *quantifiable value that defines* the point beyond which the particular limit state under investigation for a particular component will be exceeded.

Resistance can be defined in terms of load or force (static/dynamic, dead/live), stress (normal, shear, torsional), number of cycles, temperature and strain.

1. Service limit state: Relating to stress, deformation, and cracking

2. Fatigue limit state: Relating to stress range and crack growth under repetitive loads, and material toughness

3. Strength limit state: Relating to strength and stability and constructability

4. Extreme event limit state: Relating to events such as earthquakes, ice loads, vehicle and vessel collision

Building Design Criteria

The basic LRFD equation is of the form:

$$R_u \leq \phi R_n$$

R_u is required strength using LRFD load combinations shown below

R_n = nominal strength

ϕ = resistance factor

ϕR_n = design strength

Load Designations

D = dead load

L = live load

L_r = roof live load

S = snow load

R = nominal load due to initial rainwater or ice

W = wind load

E = earthquake load

Load Factors, γ, and Combinations (See Table 1)

For dead load only, D, γ_D = 1.4. For load combinations the following loads are multiplied by the appropriate factors:

TABLE 1. Load Combinations and Load Factors

Load Combination Limit State	DC, DD, D, W, EH, EV, ES, EL, PS, CR, SH	LL, IM, CE, BR, PL, LS	WA	WS	WL	FR	TU	TG	SE	EQ	IC	CT	CV
										Use One of These at a Time			
Strength I (unless noted)	γ_p	1.75	1	—	—	1	0.50/1.20	γ_{TG}	γ_{SE}	—	—	—	—
Strength II	γ_p	1.35	1	—	—	1	0.50/1.20	γ_{TG}	γ_{SE}	—	—	—	—
Strength III	γ_p	—	1	1.4	—	1	0.50/1.20	γ_{TG}	γ_{SE}	—	—	—	—
Strength IV	γ_p	—	1	—	—	1	0.50/1.20	—	—	—	—	—	—
Strength V	γ_p	1.35	1	0.4	1	1	0.50/1.20	γ_{TG}	γ_{SE}	—	—	—	—
Extreme Event I	γ_p	γ_{EQ}	1	—	—	1	—	—	—	1	—	—	—
Extreme Event II	γ_p	0.5	1	—	—	1	—	—	—	—	1	1	1
Service I	1	1	1	0.3	1	1	1.00/1.20	γ_{TG}	γ_{SE}	—	—	—	—
Service II	1	1.3	1	—	—	1	1.00/1.20	—	—	—	—	—	—
Service III	1	0.8	1	—	—	1	1.00/1.20	γ_{TG}	γ_{SE}	—	—	—	—
Service IV	1	—	1	0.7	—	1	1.00/1.20	—	1	—	—	—	—
Fatigue I— LL, IM, and CE only	—	1.5	—	—	—	—	—	—	—	—	—	—	—
Fatigue III— LL, IM, and CE only	—	0.75	—	—	—	—	—	—	—	—	—	—	—

1. $1.2D + 1.6L + 0.5(L_r \text{ or } S \text{ or } R)$, $\gamma_D = 1.2, \gamma_L = 1.6, \gamma_{Lr,S,R} = 0.5$
2. $1.2D + 1.6(L_r \text{ or } S \text{ or } R) + (0.5L \text{ or } 0.8W)$ $\gamma_D = 1.2, \gamma_L = 0.5, \gamma_{Lr,S,R} = 1.6$
3. $1.2D + 1.6W + 0.5L + 0.5(L_r \text{ or } S \text{ or } R)$ $\gamma_D = 1.2, \gamma_W = 1.6, \gamma_{Lr,S,R} = 0.5, \gamma_L = 0.5$
4. $1.2D \pm 1.0E + 0.5L + 0.2S$ $\gamma_D = 1.2, \gamma_L = 0.5, \gamma_S = 0.2, \gamma_E = 1.0$
5. $0.9D \pm (1.6W \text{ or } 1.0E)$ $\gamma_D = 0.9, \gamma_W = 1.6, \gamma_E = 1.0$

Resistance Factors, ϕ (See Table 2)

TABLE 2. Resistance Factors

Design Case	ϕ_i Value	Remarks
Flexure	$\phi_b = 0.9$	
Axial tension	$\phi_t = 0.9$	Includes tensile yielding of members and in connections
Axial compression	$\phi_t = 0.9$	
Tension rupture on net section	$\phi_t = 0.75$	Also applies to tensile strength of studs in composite components
Shear	$\phi_v = 1.0$	Without tension field action and also for shear yielding in connections
Shear	$\phi_v = 0.9$	With tension field action
Shear rupture on effective area	$\phi_{sf} = 0.75$	
Shear	$\phi_v = 0.75$	For filled and encased composite members
Shear	$\phi = 0.65$	Steel headed stud anchors in composite components
Shear	$\phi = 0.45$	Direct bond interaction
Bearing on concrete	$\phi_B = 0.65$	
Connections—fillet welds	$\phi = 0.75$	
Connections—bolts	$\phi = 0.75$	Tensile and shear strength of bolts and threaded parts and combined tension and shear in bearing type connections
Connections bolts—slip critical	$\phi = 1.00$	For standard and short-slotted holes normal to direction of load
	$\phi = 0.75$	For oversized and short slotted hole parallel to direction of load
	$\phi = 0.75$	For long slotted holes
Bearing strength at bolt holes	$\phi = 0.75$	

Bridge Design Criteria

The basic LRFD equation for bridge design is as follows:

$$\eta_D \eta_R \eta_I \sum \gamma_i Q_i \le \phi R_n$$

where
η_D = ductility factor η_R = redundancy factor η_I = operational importance factor
γ_i = load factor Φ = resistance factor Q_i = force effect
R_n = nominal resistance

Load Designations (See Table 3)
1. Permanent Loads
CR = force effects due to creep

DD = downdrag force

DC = dead load of structural components and nonstructural attachments

DW = dead load of wearing surfaces and utilities

EH = horizontal earth pressure load

EL = miscellaneous locked-in force effects resulting from the construction process, including jacking apart of cantilevers in segmental construction

ES = earth surcharge load

EV = vertical pressure from dead load of earth fill

PS = secondary forces from post-tensioning

SH = force effects due to shrinkage

2. Transient Loads
BR = vehicular braking force

CE = vehicular centrifugal force

CT = vehicular collision force

CV = vessel collision force

TABLE 3. Load Factors for Permanent Loads, γ_p

Type of Load, Foundation Type, and Method Used to Calculate Downdrag		Load Factor	
		Maximum	Minimum
DC: Component and attachments		1.25	0.90
DC: Strength IV only		1.50	0.90
DD: Downdrag	Piles, α Tomlinson method	1.40	0.25
	Piles, λ method	1.05	0.30
	Drilled shafts, O'Neill and Reese (1999) method	1.25	0.35
DW: Wearing surfaces and utilities			
EH: Horizontal earth pressure			
• Active		1.50	0.90
• At-rest		1.35	0.90
• AEP for anchored walls		1.35	N/A
EL: Locked-in construction stresses		1.00	1.00
EV: Vertical earth pressure			
• Overall stability		1.00	N/A
• Retaining walls and abutments		1.35	1.00
• Rigid buried structure		1.30	0.90
• Rigid frames		1.35	0.90
• Flexible buried structures other than metal box culverts		1.95	0.90
• Flexible metal box culverts and structural plate culverts with deep corrugations		1.50	0.90
ES: Earth surcharge		1.50	0.75

TABLE 4. Load Factors for Permanent Loads due to Superimposed Deformation, γ_p

Bridge Component	PS	CR, SH
Superstructures—segmental Concrete substructures supporting segmental Superstructures (see AASHTO section 3.12.4, 3.12.5)	1.00	See γ_p for DC on Table 3
Concrete superstructures—non-segmental	1.0	1.0
Substructures supporting non-segmental superstructures • using I_g • using $I_{effective}$	0.50 1.00	0.50 1.00
Steel substructures	1.00	1.00

EQ = earthquake load
FR = friction load
IC = ice load
IM = vehicular dynamic load allowance
LL = vehicular live load
LS = live load surcharge
PL = pedestrian live load
SE = force effect due to settlement
TG = force effect due to temperature gradient
TU = force effect due to uniform temperature
WA = water load and stream pressure
WL = wind on live load
WS = wind load on structure

Table 4 gives the Bridge Component Load Factors for Permanent Loads due to superimposed deformation.

PART 1

STRUCTURAL STEEL DESIGN

Steel Beams and Plate Girders

In the following calculation procedures, the design of steel members is executed in accordance with the load and resistance factor design (LRFD) approach of the American Institute of Steel Construction's (AISC) Steel Construction Manual 14th Edition and the American Association of State Highway Officials' (AASHTO) LRFD Bridge Design Specifications 5th Edition.

As indicated in the introduction of the LRFD, the design of the structural members will be based on the computation of the *demand* (load, Q_i) factored by the appropriate load factor(s) γ_i and the *resistance*, R_i, factored by the appropriate resistance factor(s), ϕ_i (≤ 1.0).

The notational system used conforms with that given, and it is augmented to include the following: A_w = area of flange, in.² (cm²); A_w = area of web, in.² (cm²); b_f = width of flange, in. (mm); d = depth of section, in. (mm); d_w = depth of web, in. (mm); t_f = thickness of flange, in. (mm); t_w = thickness of web, in. (mm); L' = unbraced length of compression flange, in. (mm); F_y = yield-point stress, psi (kPa).

MOST ECONOMIC SECTION FOR A BEAM WITH A CONTINUOUS LATERAL SUPPORT UNDER A UNIFORM LOAD

A beam on a simple span of 30 ft (9.2 m) carries a uniform live load (L) of 1650 lb/lin ft (24,079.9 N/m). The compression flange is laterally supported along its entire length. Select the most economic ASTM A992 section.

Calculation Procedure:

1. Compute the factored maximum bending moment (Demand)
Assume that the beam weighs, (dead load, D) 50 lb/lin ft (729.7 N/m) and satisfies the requirements of a compact section as set forth in the *Specification*.

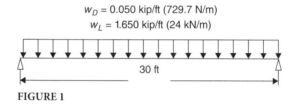

FIGURE 1

ASTM A992 steel implies that F_y = 50 ksi (345 MPa) and F_u = 65 ksi (448 MPa) according to Table 2-4 of the AISC Manual.

$$w_u = (\gamma_D w_D + \gamma_L w_L) \qquad (1)$$

$\gamma_D = 1.2$ and $\gamma_L = 1.6$

$w_u = 1.2(0.050 \text{ kip/ft}) + 1.6(1.650 \text{ kips/ft}) = 2.70 \text{ kips/ft (39 kN/m)}$

$$M_u = \frac{2.70 \text{ kips/ft } (30 \text{ ft})^2}{8} = 303.75 \text{ kip·ft } (412 \text{ kN · m}) = \Sigma \gamma_i Q_i$$

The beam is compact and continuously braced, therefore the yielding limit state controls and $M_n = M_p$.

2. Determine moment resistance, ϕM_r, and select the most economic section
From Table 3-2 of the AISC Manual Select Section W16 × 45 with Z_x = 82.3 in.³ (1,348,655 mm⁴) and $\phi_b M_{px}$ = 309 kip · ft (419 kN · m) for F_y = 50 ksi (345 MPa).

3. Compare load (demand and resistance)
Factored load, R_u = 303.75 kip · ft (412 kN · m) is < factored resistance, $\phi R_r = \phi_b M_{px}$ = 309 kip · ft (419 kN · m).

Therefore section is OK.

MOST ECONOMIC SECTION FOR A BEAM WITH AN INTERMITTENT LATERAL SUPPORT UNDER A UNIFORM LOAD

A beam on a simple span of 25 ft (7.6 m) carries a total uniformly distributed live load of 45 kips (200.2 kN), and the weight of the beam is 50 lb/ft. The member is laterally supported at 5-ft (1.5-m) intervals. Select the most economic member using A992 steel.

Calculation Procedure:

1. Compute the factored maximum bending moment (Demand)
The beam weighs, (dead load, D) 50 lb/lin ft (729.7 N/m) and satisfies the requirements of a compact section as set forth in the *Specification*. $L = 45$ kips/25 = 1.80 kips/ft (26 kN/m)

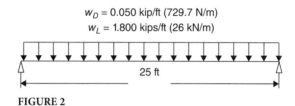

FIGURE 2

ASTM A992 steel implies that $F_y = 50$ ksi (345 MPa) and $F_u = 65$ ksi (448 MPa) according to Table 2-4 of the AISC *Manual*.

$$w_u = (\gamma_D w_D + \gamma_L w_L) \tag{1}$$

$\gamma_D = 1.2$ and $\gamma_L = 1.6$

$w_u = 1.2(0.050 \text{ kip/ft}) + 1.6(1.800 \text{ kips/ft}) = 2.94$ kips/ft (43 kN/m)

$$M_u = \frac{2.94 \text{ kips/ft } (25 \text{ ft})^2}{8} = 229.69 \text{ kip} \cdot \text{ft} = \Sigma \gamma_i Q_i = R_u = 229.69 \text{ kip} \cdot \text{ft (3.11 N} \cdot \text{m)}$$

2. Compute Moment Resistance, ϕR_r, and select the most economic section
Unbraced length, $L_u = 5$ ft (1.5 m). From Table 3-1 of the AISC *Manual*, a uniformly loaded beam braced at fifth points (5/25 = 1/5 points), $C_b = 1.00$ at the center of the beam. C_b = lateral torsional buckling modification factor.
From AISC *Manual* Table 3-2 Try Member W10 × 54 with moment resistance $\phi_b M_p = 250$ kip · ft (339 N · m). Refer to Table 3-10 of the AISC *Manual*. The intersection of the curve for W10 × 54 for an unbraced length of fifth (1.5 m) corresponds to 250 kip · ft (339 N · m) since $C_b = 1.0$.

3. Select the most economic section
Since section W10 × 54 with moment resistance $\phi_b M_p$ of 250 kip · ft (339 N · m) is greater than the required strength R_u of 229.69 kip · ft (311 N · m), select W10 × 54.

4. Compare load (demand) and resistance
Factored load, $\Sigma \gamma Q = 229.68$ kip · ft (311 N · m) is < factored resistance, $\phi R_r = \phi_b M_{px} = 250$ kip · ft (339 N · m).
Therefore section is OK.

STRUCTURAL STEEL DESIGN 1.11

DESIGN OF A BEAM WITH REDUCED MOMENT RESISTANCE

The compression flange of the beam in Fig. 1a will be braced only at points A, B, C, D, and E. Using AISC data, a designer has selected W21 × 55 section for the beam. Verify the design. Assume all loads are dead loads.

Calculation Procedure:

1. Calculate the reactions; construct the shear and bending-moment diagrams
The results of this step are shown in Fig. 3.

2. Determine unbraced lengths, L_b
L_b is the length between braced points of the compression flange of a beam, i.e., unbraced length. For Segment AB (Overhang but braced), $L_{b\text{-}AB}$ = 8.0 ft (2.44 m); segment BC, $L_{b\text{-}BC}$ = 15.0 ft (4.57 m); segment CD, $L_{b\text{-}CD}$ = 16.5 ft (5 m); and for segment DE, $L_{b\text{-}DE}$ = 8.5 ft (2.6 m).

3. Record properties of selected section from the Tables in AISC Manual
For member W21 × 55, A = 16.2 in.^{2} (10,452 mm^2), d = 20.8 in. (528 mm), t_w = 0.375 (9.5 mm), b_f = 8.22 in. (209 mm), t_f = 0.522 (13.3 mm), I_x = 1140 in.^{4} (47,450 cm^4), S_x = 110 in.^{3} (1802 cm^3), r_x = 8.40 in. (213.4 mm), Z_x = 126 in.^{3} (2064 cm^3), I_y = 48.4 in.^{4} (2014 cm^4), S_y = 11.8 in.^{3} (193 cm^3), r_y = 1.73 in. (43.9 mm), Z_y = 18.4 in.^{3} (301 cm^3), J = 1.24 in.^{4} (51.6 cm^4), C_w = 4980 in.^{6} (1,337,308 cm^6).

4. Compute limiting unbraced length, L_p, L_r

$$L_p = 1.76 r_y \sqrt{\frac{E}{F_y}} \qquad \text{(AISC Eq. F2-5)}$$

L_p = 6.11 ft (1.86 m)

$$L_r = 1.95 r_{ts} \frac{E}{0.7 F_y} \sqrt{\frac{Jc}{S_x h_o} + \sqrt{\left(\frac{Jc}{S_x h_o}\right)^2 + 6.76\left(\frac{0.7 F_y}{E}\right)^2}} \qquad \text{(AISC Eq. F2-6)}$$

L_r = 17.40 ft (5.3 m)

where
  E = modulus of elasticity of steel = 29,000 ksi (200,000 MPa)
  J = torsional constant in.^{4} (mm^4)
  S_x = elastic section modulus taken about the x-axis in.^{3} (mm^3)
  h_o = distance between flange centroids, in. (mm)
  F_y = specified yield stress of steel = 50 ksi (345 MPa)

$$r_{st}^2 = \frac{\sqrt{I_y C_w}}{S_x} \qquad \text{(AISC Eq. F2-7)}$$

And the coefficient c = 1 for doubly symmetric I shapes

5. Determine Nonuniform Moment Modification Factor, C_b
Generally the value of C_b is dependent on whether the moment is uniform or nonuniform between braced points. For uniform (constant) moment between braced points and for non-braced overhang (cantilever members, C_b = 1.0). For nonuniform moment between braced points, Eq. (F1-1) from AISC Specifications should be used in computing C_b as follows:

$$C_b = \frac{12.5 M_{max}}{2.5 M_{max} + 3 M_A + 4 M_B + 3 M_C}$$

where

M_{max} = absolute value of maximum moment in the unbraced segment, kip-in. (N-mm)
M_A = absolute value of moment at ¼ point of the unbraced segment, kip-in. (N-mm)
M_B = absolute value of moment at midpoint of the unbraced segment, kip-in. (N-mm)
M_C = absolute value of moment at ¾ point of the unbraced segment, kip-in. (N-mm)

From Table 5 segment CD governs with the lowest value of $C_b = 1.03$. This will provide the lowest moment capacity or resistance.
Since $L_{b\text{-}CD} = 16.5$ ft (5 m) $> L_p = 6.11$ ft (1.86 m) but $< L_r = 17.40$ ft (5.3 m).
Therefore nominal moment resistance, M_n is given by AISC Eq. (F2-2) as

$$M_n = C_b \left[M_p - \left(M_p - 0.7 F_y S_x\right)\left(\frac{L_b - L_p}{L_r - L_p}\right) \right] \leq M_p \qquad \text{(AISC Eq. F2-2)}$$

Equation (F2-2) gives $M_P = Z_x F_y = 50$ ksi $\times 126$ in.$^3 = 6300$ kip-in. $= 525$ kip · ft (712 kN · m)
$M_n = 1.03[6300 - \{6300 - 0.7 \times 50 \times 110 \times (16.5 \times 12 - 73.33)/(208.82 - 73.33)\}] = 4181.5$ kip-in. $= 384.46$ kip · ft (472 kN · m) $< M_p = 522$ kip · ft (712 kN · m)
The maximum factored moment is $M_u (\gamma R_i) = 1.2 \times 148$ kip · ft $= 177.6$ kip · ft (241 kN · m)
$\phi_b M_n = \phi_b M_r = 0.9 \times 348.46$ kip · ft $= 313.61$ kip · ft (425 kN · m)
Since 177.6 kip · ft (241 kN · m) $(\gamma R_i) < 313.61$ kip · ft (425 kN · m) $(\phi_b M_r)$, design is OK.
In Summary, If $L_b \leq L_p$, then $M_n = M_p = Z_x F_y$, if $L_p < L_b \leq L_r$, then

$$M_n = C_b \left[M_p - \left(M_p - 0.7 F_y S_x\right)\left(\frac{L_b - L_p}{L_r - L_p}\right) \right] \leq M_p$$

and if $L_b > L_r$, then $M_n = F_{cr} S_x$, where

$$F_{cr} = \frac{C_b \pi^2 E}{\left(\frac{L_b}{r_{ts}}\right)^2} \sqrt{1 + 0.078 \frac{Jc}{S_x h_o}\left(\frac{L_b}{r_{ts}}\right)^2}$$

TABLE 5. Computation of C_b (Moments are in kip · ft)

		¼ pt.	Mid-point	¾ pt.		Ratios			
Segment	M_{max}	M_A	M_B	M_C	M_{max}	M_A/M_{max}	M_B/M_{max}	M_C/M_{max}	C_b
A-B	148	19	50	93	1	0.128	0.338	0.628	2.04
B-C	148	66.2	0	55.3	1	0.447	0.000	0.374	2.52
C-D	133	122	133	126	1	0.917	1.000	0.947	1.03
D-E	102	83.6	60.2	32.4	1	0.820	0.590	0.318	1.51

TABLE 5(a). Moments in SI Units (kN · m)

Segment	M_{max}	M_A	M_B	M_C
A-B	201	25.8	67.8	126
B-C	201	89.8	0.00	75.0
C-D	180	165	180	171
D-E	138	113	81.6	43.9

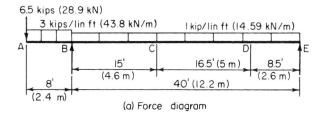
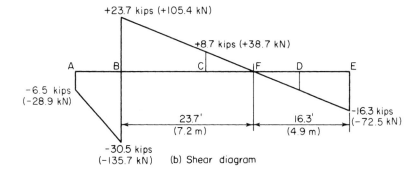
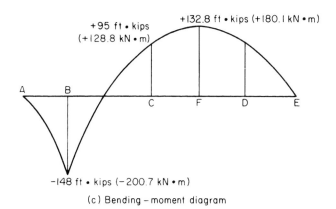

FIGURE 3

DESIGN OF A COVER-PLATED BEAM

Following the fabrication of a W18 × 40 beam, a revision was made in the architectural plans, and the member must now be designed to support the loads shown in Fig. 3a. Cover plates are to be welded to both flanges to develop the required strength. Design these plates and their connection to the W shape, using fillet welds class E70 series electrodes. The member has continuous lateral support. Assume built-up section will be compact.

Calculation Procedure:

1. Construct the shear and bending-moment diagrams
These are shown in Fig. 4. Also, $M_E = 340.3$ ft · kips (461 kN · m). Assuming all loads are permanents loads, i.e., $\gamma_D = 1.2$, then the required strength, $R_u = 1.2 \times 340.3$ kip · ft = 408.36 kip · ft (553.6 kN · m).

2. Record the properties of the beam section
Refer to the AISC *Manual*, and record the following properties for the W18 × 40; $d = 17.9$ in. (455 mm); $b_f = 6.02$ in. (153 mm); $t_f = 0.525$ in. (13.3 mm); $I = 612$ in.^{4} (25,473 cm^4); $Z_x = 78.4$ in.^{3} (1285 cm^3). $S_x = 68.4$ in.^{3} (1120 cm^3).

3. Compute moment capacity/available strength of W18 × 40
Since the built-up section is assumed compact and continuously laterally braced, the nominal moment is equal to the plastic moment, $M_n = M_p = Z_x F_y = 78.4$ in.^{3} × 50 ksi = 3920 kip-in. = 326.7 kip · ft (442.8 kN · m). Therefore available strength is $\phi M_n = \phi M_p = 9 \times 326.7$ kip · ft = 294 kip · ft (398.6 kN · m). This is less than the required strength [294 kip · ft (398.6 kN · m) < 408.36 kip · ft (553.6 kN · m)], therefore cover plates are needed.

4. Determine the plastic section modulus required, Z_{reqd}
The required strength R_u or $M_u = 408.36$ kip · ft (553.6 kN · m). Equating required strength to available strength, R_r, which in this case equals $\phi M_n = \phi M_p = \phi_b Z_x F_y$, i.e., $M_u = \phi_b Z_x F_y$, it implies that the required plastic section modulus, $Z_{reqd} = M_u / \phi_b F_y = (408.36$ kip · ft × 12 ft/in.)/ $(0.9 \times 50$ ksi) = 108.896 in.^{3} (1784 cm^3). However, the required plastic section modulus, $Z_{reqd} = Z_{W\text{-shape}} + Z_{coverplates}$. $Z_{W18 \times 40} = 78.4$ in.^{3} $Z_{coverplates} = 2A_p (d/2 + t_p/2)$. Assume thickness of one cover plate $t_p = 0.375$ in. (⅜) (9.5 mm). d = depth of W section = 17.9 in. (455 mm). The required area, A_p, of one cover plate is given by $A_p = (Z_{reqd} - Z_w)/(d + t_p) = (108.896 - 78.4)/(17.9 + 375) = 30.496/18.275 = 1.69$ in.^{2} (10.8 cm^2). Width of cover plate required, $b_p = 1.69/0.375 = 4.5$ in. (113 mm). Use a 5 in. × ⅜ in. plate.

5. Check width-to-thickness (b/t) ratio of cover plate
From AISC *Manual* 14th edition Table B4.1b, the b/t ratio of cover plates in compression due to flexure shall be as follows:

$$\frac{b}{t} \leq 1.12\sqrt{\frac{E}{F_y}} \text{ for compact sections or } \leq 1.40\sqrt{\frac{E}{F_y}} \text{ for noncompact sections}$$

$b_t = 5.0/0.375 = 13.33$ and $1.12 \times (29{,}000 \text{ ksi}/50 \text{ ksi})^{\wedge}.5 = 26.97$. Therefore $b/t \leq 1.12\sqrt{E/F_y}$. Hence OK.

6. Check adequacy of beam and cover plate section

$$Z_{provided} = Z_w + Z_{cover\ plates}$$

$Z_{cover\ plates} = 2A_p (d/2 + t_p/2) = 2 \times (5 \text{ in.} \times 0.375 \text{ in.}) \times (17.9/2 + 0.375/2) = 34.265$ in.^{3}; $Z_w = 78.4$ in.^{3} Therefore $Z_{provided} = 78.3 + 34.265 = 112.66$ in.^{3} > Z_{reqd} (108.896 in.^{3}). Available strength provided, $\phi_b M_n = \phi_b Z_x F_y = 0.9 \times 112.66 \times 50 = 5070$ kip-in. = (422.5 kip · ft (572.8 kN · m)) > $M_u = 408.36$ kip · ft (553.6 kN · m).

Therefore modified section is OK.

a. Locate the points where the cover plates are not needed
Compare the nominal moment capacity M_n of the W shape without the cover plates to the unfactored maximum moment, M_{max} on the beam. From Fig. 4 the maximum moment is 340.3 kip · ft (461.4 kN · m) and the nominal moment capacity of the W section is 326.7 kip · ft (442.8 kN · m). The segment where the moment exceeds 326.7 kip · ft (442.8 kN · m) on both sides of the maximum moment will require cover plates.

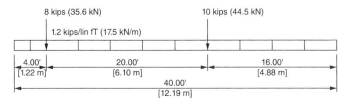

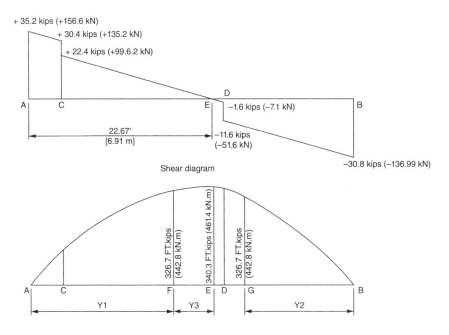

FIGURE 4A. Bending moment and shear force diagrams.

b. Locate points at which computed moments occur
These points are F and G (Fig. 4). Thus, $M_F = 35.2 y_1 - 8(y_1 - 4) - \frac{1}{2}(1.2 y_1^2) = 326.7$ kip · ft; $y_1 = 9.03$ ft (2.75 m); $M_G = 30.8 y_2 - \frac{1}{2}(1.2 y_2^2) = 326.7$; $y_2 = 9.15$ ft (2.79 m). For symmetry, center the cover plates about mid-span, placing the theoretical cutoff points at 9 ft 3 in. (2.82 m) from each support.

7. Compute axial force in the cover plate
Calculate the axial force P kips (N) in the cover plate at its end by computing the mid-thickness bending stress at the cover plate. Thus $\sigma_{bending} = My/I = 326.7(12)(9.325)/925.15 = 39.52$ ksi (272 MPa). Then $P = A_{cover\ plate} \times \sigma_{bending} = 1.875(39.52) = 74.1$ kips (339,004 N). Determine the length of fillet weld required to transmit this force to the W shape. Use a ¼-in. (6.35-mm) fillet weld, which satisfies the requirements of the specification. The nominal capacity, $R_n = F_{ne}A_{we}$, where $F_{ne} = 0.60 F_{EXX}$ and A_{we} = Effective area of weld = weld size/$\sqrt{2}$. Therefore $R_n = 0.60(70\ \text{ksi}) \times (\frac{1}{4})/(2^{1/2}) = 7.42$ kips/in. Length of weld $L_{weld} = P/\phi R_n = 74.1/(0.75 \times 7.42) = 13.3$ in. = 6.65 in. of weld on one side of cover plate. Use 7 in. long weld on each side of cover plate.

8. Extend cover plates

In accordance with the Specification, extend the cover plates 20 in. (508.0 mm) beyond the theoretical cutoff point at each end, and supply a continuous ¼-in. fillet weld along both edges in this extension. This requirement yields 27 in. (686 mm) of weld as compared with the 7 in. (178 mm) needed to develop the plate.

	Width	Thickness	A	y	Ay	$K(y-y_{NA})$	AK^2	I_o	I_{xx}
Top cover plate	5	0.375	1.88	18.46	34.62	9.14	156.55	0.02	156.57
W18 × 40			17.9	11.80	9.33 110.04	0.00	0.00	612.00	612.00
Bottom cover plate	5	0.375	1.88	0.19	0.35	−9.14	145.55	0.02	156.57
			15.55		145.00				925.15

$$y_{NA} = 9.325 \text{ in.} \qquad S_{top} = 99.21 \text{ in.}^3$$

9. Calculate horizontal shear flow at the inner surface of the cover plate

Choose F or G, whichever is larger. Design the intermittent fillet weld to resist this shear flow. Thus $V_F = 35.2 − 8 − 1.2(9.03) = 16.36$ kips (72.66 kN); $V_G = −30.8 + 1.2(9.25) = −19.7$ kips (−87.47 kN). Then $q = VQ/I = 1.2 \times 19,700(1.875)(9.325)/925.15 = 446.77$ lb/in. (78,241.42 N/m). The *Specification* calls for a minimum weld length of 1.5 in. (38.10 mm). Let s denote the center-to-center spacing as governed by shear. Then $s = 2(1.5)(7420)/446.77 = 50$ in. (411.48 mm). However, the AWS *Specification* imposes additional restrictions on the intermittent weld spacing. To preclude the possibility of error in fabrication, provide an identical spacing at the top and bottom as 24 times the thickness of the thinner part. Thus, $S_{max} = 24(0.375) = 9$ in. (228.6 mm). Therefore, use a ¼-in. (6.35-mm) fillet weld, 1.5 in. (38.10 mm) long, and 9 in. (228.6 mm) on centers, as shown in Fig. 4b.

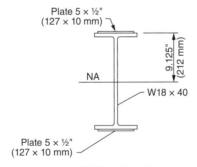

Reinforced section

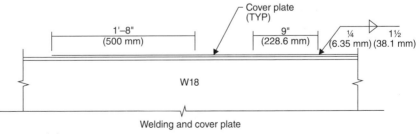

Welding and cover plate

FIGURE 4(B)

DESIGN OF A CONTINUOUS BEAM

The beam in Fig. 5a is continuous from A to D and is laterally supported at 5-ft (1.5-m) intervals. Design the member. Assume all loads are live loads.

Calculation Procedure:

1. Find the bending moments at the interior supports; calculate the reactions and construct shear and bending-moment diagrams

The maximum moments are +101.7 ft · kips (137.9 kN · m) and −130.2 ft · kips (176.55 kN · m). The factored moments M_u are 1.6 × 101.7 kip · ft = 162.72 kip · ft (220.6 kN · m) and 1.6 × (−130.2) kip · ft = 208.32 kip · ft (282.44 kN · m), 1.6 × (−115.9) kip · ft = 185.44 kip · ft (251.42 kN · m).

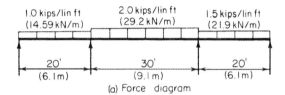

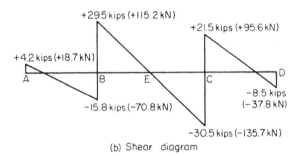

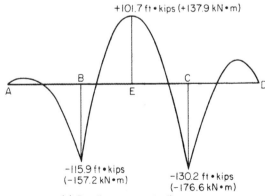

FIGURE 5

1.18 STRUCTURAL ENGINEERING

2. Calculate the modified maximum moments
Calculate these moments in the manner prescribed in the AISC *Specification*. The clause covering this calculation is based on the postelastic behavior of a continuous beam. (Refer to a later calculation procedure for an analysis of this behavior.) Modified maximum moments: $+162.72 + 0.1(0.5)(185.44 + 208.32) = +182.408$ ft · kips (247.3 kN · m); $0.9(-208.32) = -187.5$ ft · kips (-254.2 kN · m); design moment = 187.5 ft · kips (254.2 kN · m).

3. Select the beam size
Thus, $Z = M/\phi F_y = 187.5(12)/(0.9 \times 50) = 50$ in.^{3} (819.35 cm^3). Use W12 × 35 with $Z = 51.2$ in.^{3} (839.02 cm^3); $L_p = 5.44$ ft (1.66 m).

SHEARING STRESS IN A BEAM—EXACT METHOD

Calculate the maximum shearing stress in a W18 × 55 beam at a section where the vertical shear is 70 kips (311.4 kN).

Calculation Procedure:

1. Record the relevant properties of the member
The shearing stress is a maximum at the centroidal axis and is given by $v = VQ/(It)$. The static moment of the area above this axis is found by applying the properties of the WT9 × 27.5, which are presented in the AISC *Manual*. Note that the T-Section considered is one-half the wide-flange section being used. See Fig. 6. The properties of these sections are $I_{W18 \times 55} = 1140$ in.^{4} (47,450 cm^4); $A_T = 8.10$ in.^{2} (52.261 cm^2); $t_w = 0.39$ in. (9.906 mm); $y_m = 9.06 - 2.16 = 6.90$ in. (175.26 mm).

2. Calculate the shearing stress at the centroidal axis
Substituting gives $Q = 8.10(6.90) = 55.9$ in.^{3} (916.20 cm^3); then $v = 70,000(55.9)/[1140(0.39)] = 8,801.2$ lb/in.^{2} (60.68 MPa).

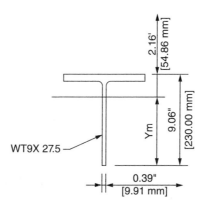

FIGURE 6

SHEARING STRESS IN A BEAM: APPROXIMATE METHOD

Solve the previous calculation procedure, using the approximate method of determining the shearing stress in a beam.

Calculation Procedure:

1. Assume that the vertical shear is resisted solely by the web
Consider the web as extending the full depth of the section and the shearing stress as uniform across the web. Compare the results obtained by the exact and the approximate methods.

2. Compute the shear stress
Take the depth of the web as 18.12 in. (460.248 mm), $v = 70,000/[18.12(0.39)] = 9910$ lb/in.² (68.3 MPa). Thus, the ratio of the computed stresses is $8801.2/9910 = 0.89$. Since the error inherent in the approximate method is not unduly large, this method is applied in assessing the shear capacity of a beam. The available shear strength ϕV_n for each rolled section is recorded in Tables 3-2, 3-6 to 3-9 of the AISC *Manual*.

The design of a rolled section is governed by the shearing stress only in those instances where the ratio of maximum shear to maximum moment is extraordinarily large. This condition exists in a heavily loaded short-span beam and a beam that carries a large concentrated load near its support.

MOMENT CAPACITY OF A WELDED PLATE GIRDER

A welded plate girder is composed of a 66- × ⅜-in. (1676.4 × 9.53-mm) web plate and two 20 × ¾-in. (508.0 × 19.05-mm) flange plates. The unbraced length of the compression flange is 18 ft (5.5 m).
If $C_b = 1$, what bending moment can this member resist?

Calculation Procedure:

1. Compute section properties of welded plate girder

	Width	Thickness	A	Y	Ay	$K(y-y_{NA})$	AK^2	Io	I_{xx}
Top flange	20	0.75	15.000	67.13	1006.88	33.38	16708.36	0.70	16709.06
Web	66	0.375	24.75	33.75	835.31	0.00	0.00	8984.2	8984.25
Bottom flange	20	0.75	15.00	0.38	5.63	33.38	16708.3	0.70	16709.0
			54.75		1847.81			8985.6	42402.38

$y_{NA} = 33.75$ in. $S_{bot} = 1256.37$ in.³ $S_{top} = 1256.37$ in.³ $h_o = 66.75$ in.

J	x	Ax	k_x	$A_x K_x^2$	I_y
2.81	10	150	0	0	500
1.160	10	247.5	0	0	0.290
2.81	10	150	0	0	500
6.79		547.5			1000.29

$X_{na} = 10$ in. $r_y = 4.27$ in.

2. Check compactness of girder section
From AISC Table 4.1b Cases 11 and 15, for flange compactness, $b/t = 10/.75 = 13.33 \le \lambda_p = 0.38(E/F_y)^{0.5} = 9.15$. Therefore flanges are not compact. Check flange slenderness, $\lambda_r = 0.95 \times (k_c E/F_L)^{0.5}$, $k_c = 4/(h/t_w)^{0.5} = 4/(66/.375)^{0.5} = 0.30 < 0.35$ therefore use $k_c = 0.35$. Since $S_{xt} = S_{xc} = 1$, then $F_L = 0.70 F_y = 0.70 \times 50 = 35$ ksi. $\lambda_r = 0.95(.35 \times 29000/35)^{0.5} = 16.18$ which is >13.33. Check web compactness and/or slenderness, $h/t_w = 66/0.375 = 176$. For compactness $h/t_w \le 3.76(E/F_y)^{0.5} = 3.76 \times (29,000/50)^{0.5} = 90.55$. Therefore web is not compact. But, $5.70 \times (29,000/50)^{0.5} = 137.3 < 176$. Hence web is slender. Therefore the plate girder section is noncompact. AISC section F4 applies.

3. Check unbraced length limit
The given unbraced length $L_b = 18$ ft (5.49 m). First compute L_p = the unbraced length for the limit state of yielding $= 1.1 \times r_t \times e\,(E/F_y)$, r_t is the effective radius of gyration for lateral torsional buckling determined as follows:

$$r_t = \frac{b_{fc}}{\sqrt{12\left(\dfrac{h_o}{d} + \dfrac{1}{6}a_w \dfrac{h^2}{h_o d}\right)}}$$

b_{fc} = width of compression flange = 20 in., t_{fc} = thickness of compression flange = 0.75 in. $a_w = (h_c t_w / b_{fc} t_{fc}) = (66 \times 0.375/(20 \times 0.75) = 1.65$, where $h_c = h = 66$, i.e., twice the distance from the centroidal axis to the inside face of the compression flange for a welded plate girder. $d = 66 + 1.5 = 67.5$ in., $h_o = 66.75$ as previously computed. Therefore $r_t = 20/[12 \times (66.75/67.5 + 1/6 \times (1.65) \times 66^2/(66.75 \times 67.5))]^{0.5}$ $r_t = 5.15$ in. Therefore $L_p = 1.1 \times 5.15 \times (29,000/50)^{0.5} = 136.4$ in. (3.5 m) = 11.4 ft. Therefore $L_b > L_p$. Now check if L_b is less than L_r.

$$L_r = 1.95 r_t \frac{E}{F_L} \sqrt{\frac{J}{S_{xc} h_o} + \sqrt{\left(\frac{J}{S_{xc} h_o}\right)^2 + 6.76\left(\frac{F_L}{E}\right)^2}}$$

$$L_r = 1.95(5.15)\frac{29000}{35}\sqrt{\frac{6.76}{1256.37 \times 66.75} + \sqrt{\left(\frac{6.79}{1256.37 \times 66.75}\right)^2 + 6.76\left(\frac{35}{29000}\right)^2}} = 330.5 \text{ in.}$$

$L_r = 27.5$ ft (8.4 m). Hence L_b (18 ft) $< L_r$. Since $L_p < L_b < L_r$, nominal moment capacity, M_n, will be determined as follows:

$$M_n = C_b\left[R_{pc}M_{yc} - (R_{pc}M_{yc} - F_L S_{xc})\left(\frac{L_b - L_p}{L_r - L_p}\right)\right] \le R_{pc}M_{yc}, M_{yc} = S_{xc}F_y,$$

$M_{yc} = 1256.37 \times 50 = 62,818.5$ kip-in (7097 kN · m); R_{pc} is web plastification factor and is dependent on the ratio of I_{yc}/I_y and web slenderness, λ. $I_{yc}/I_y = 500/1000.29 = 0.5 > 0.23$, and $h_c/t_w > \lambda_{pw}$, i.e., $176 > 90.55$, therefore

$$R_{pc} = \left[\frac{M_p}{M_{yc}} - \left(\frac{M_p}{M_{yc}} - 1\right)\left(\frac{\lambda - \lambda_{pw}}{\lambda_r - \lambda_{pw}}\right)\right] \le \frac{M_p}{M_{yc}},$$

$M_p = Z_x F_y$, $Z_x = Z_{\text{flanges}} + Z_{\text{web}} = 2A_f(h/2 + t_f/2) + 2[(h/2)(t_w)(h/4)] = 2 \times (20 \times 0.75)(66/2 + 0.75/2) = 1001.25$ in.^{4} $M_p = 1001.25 \times 50 = 50,062.5$ kip-in (5656.25 kN · m). $M_p/M_{yc} = 62,818.5/50,062.5 = 1.25$. $(\lambda - \lambda_{pw}/\lambda_r - \lambda_{pw}) = [(176 - 90.55)/(137.3 - 90.55)] = 1.83$. Therefore, $R_{pc} = 1.25 -$

$(1.25 - 1)(1.83) = 0.79$. Hence, $M_n = 1.0[0.79 \times 50,062.5 - (0.79 \times 50,062.5 - 35 \times 35 \times 1256.37) \times [18 - 11.4/(27.5 - 11.4)] = 37,735.98$ kip-in (4263.6 kN · m).

The available strength is $\phi M_n = 0.9 \times 37,735.98 = 33,962.39$ kip-in (3837.23 kN · m) = 2830.2 kip · ft.

The bending moment that can be resisted by this member is 2830.2 kip · ft.

ANALYSIS OF A RIVETED PLATE GIRDER

A plate girder is composed of one web plate 48 × ⅜ in. (1219.2 × 9.53 mm); four flange angles 6 × 4 × ¾ in. (152.4 × 101.6 × 19.05 mm); two cover plates 14 × ½ in. (355.6 × 12.7 mm). The flange angles are set 48.5 in. (1231.90 mm) back to back with their 6-in. (152.4-mm) legs outstanding; they are connected to the web plate by ⅞-in. (22.2-mm) rivets. If the member has continuous lateral support, what bending moment may be applied? What spacing of flange-to-web rivets is required in a panel where the vertical shear is 180 kips (800.6 kN)?

Calculation Procedure:

1. Obtain the properties of the angles from the AISC Manual
Record the angle dimensions as shown in Fig. 7.

2. Check the cover plates and angles for compactness
AISC Table 4.1(b) For cover plate compactness, $b/t < 1.12eE/F_y$. $b = 14/0.5 = 28$, $1.12eE/F_y = 1.12 \times (e\,(29000/50)) = 26.97 < 28$. But $1.40eE/F_y = 33.7 > 28$, therefore cover plates are not slender. For angles, $b/t < 0.54eE/F_y$. $b/t = 6/0.75 = 8.0$, $0.54eE/F_y = 13.0 > 8.0$. Therefore angles are compact. Check the web for compactness h/t_w, $< 3.76\ eE/Fy$, $48/.375 = 128$ and $3.76\ eE/Fy = 90.55 < 128$. And $5.70\ eE/F_y = 137.27 > 128$. Therefore web is noncompact but not slender. Section is therefore noncompact but not slender.

3. Compute the gross flange area and rivet-hole area
Ascertain whether the *Specification* requires a reduction in the flange area. Therefore gross flange area = $2(6.94) + 7.0 = 20.88$ in.² (134.718 cm²); area of rivet holes = $2(½)(1) + 4(¾)(1) = 4.00$ in.² (25.808 cm²); allowable area of holes = $0.15(20.88) = 3.13$. The excess area = hole area – allowable area = $4.00 - 3.13 = 0.87$ in.² (5.613 cm²). Consider that this excess area is removed from the outstanding legs of the angles, at both the top and the bottom.

4. Compute the moment of inertia of the net section

	in.⁴	dm⁴
One web plate, I_0	3,456	14.384
Four flange angles, I_0	35	0.1456
$Ay^2 = 4(6.94)(23.17)^2$	14,900	62.0184
Two cover plates:		
$Ay^2 = 2(7.0)(24.50)^2$	8,400	34.9634
I of gross section	26,791	111.5123
Deduct $2(0.87)(23.88)^2$ for excess	991	4.12485
area I of net section	25,800	107.387

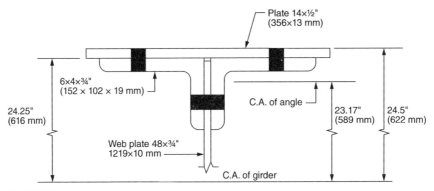

FIGURE 7

5. Compute the nominal moment capacity of the net section of girder section

The section is a braced noncompact section that is not slender, therefore nominal moment capacity, M_n will be given by $M_n = M_p - (M_p - 0.7 F_y S_x)[\lambda - \lambda_{pf}/\lambda_{rf} - \lambda_{pf}]$ from AISC Eq. (F3-1), where $\lambda_{pf} = \lambda_p$ is the limiting slenderness for a compact flange and $\lambda_{pr} = \lambda_r$ is the limiting slenderness for a non-compact flange. $M_p = Z F_y$, $Z_{girder} = Z_{cover\,plate} + Z_{angle} + Z_{web} = 2 A_{cover\,plate}\,(h_{coverplate}/2) + 2 A_{angles}\,(h_{angles}/2) + 2[(h_{web}/2)\,(t_w)\,(h_{web}/4)] = 2 \times 7 \times (24.5) + 2 \times 6.94 \times 23.17 + 2 \times 24 \times .375 \times 12 = 880.6$ in.³. Therefore $M_p = 880.6 \times 50 = 44{,}029.98$ kip-in = 3,669.17 kip · ft (4974.7 kN · m). Compute section Modulus to compression flange, $S_{xc} = 1080.8$ in³ (17.71 dm³). $\lambda = b_f/2t_f = 14/(0.5) = 28$. λ_p 26.97 and $\lambda_r = 33.7$. Therefore $M_n = 44{,}029.98\,(44{,}029.98 - 0.7 \times 50 \times 1080.8)\,[(28 - 26.97)/(33.7 - 26.97)] = 43{,}080.79$ kip-in. = 3590 kip · ft (4867.47 kN · m). The available bending strength is $\phi M_n = 0.9 \times 3590$ kip · ft = 3231 kip · ft (4380.73 kN · m).

6. Calculate the horizontal shear flow to be resisted

Here Q of flange = $13.88(23.17) + 7.0(24.50) - 0.87(23.88) = 472$ in.³ (7736.1 cm³); $q = VQ/I = 180{,}000(472)/25{,}800 = 3290$ lb/in (576,167.2 N/m). From a previous calculation procedure, $R_{ds} = 18{,}1040$ lb (80,241.9 N); $R_b = 42{,}440(0.375) = 15{,}900$ lb = (70,723.2 N); $s = 15{,}900/3290 = 4.8$ in. (121.92 mm), where s = allowable rivet spacing, in. (mm). Therefore, use a 4¾-in. (120.65-mm) rivet pitch. This satisfies the requirements of the *Specification. Note:* To determine the allowable rivet spacing, divide the horizontal shear flow into the rivet capacity.

DESIGN OF A WELDED PLATE GIRDER

A plate girder of welded construction is to support the loads shown in Fig. 8a. The distributed load will be applied to the top flange, thereby offering continuous lateral support. At its ends, the girder will bear on masonry buttresses. The total depth of the girder is restricted to approximately 70 in. (1778.0 mm). Select the cross section, establish the spacing of the transverse stiffeners, and design both the intermediate stiffeners and the bearing stiffeners at the supports. Assume all loads are live loads and $F_y = 36$ ksi.

Calculation Procedure:

1. Construct shear and bending moment diagrams

These diagrams are shown in Fig. 8.

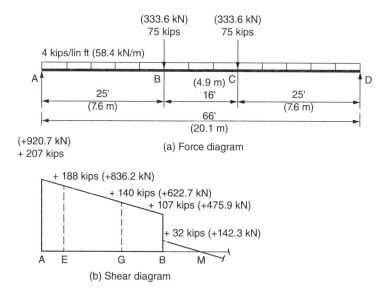

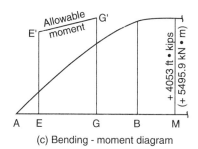

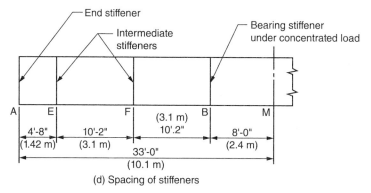

FIGURE 8

2. Choose the web-plate dimensions
Since the total depth is limited to about 70 in. (1778.0 mm), use a 68-in. (1727.2-mm) deep web plate. Determine the plate thickness, using Table B4.1b of the AISC *Specification* limits, which are a slenderness ratio $h/t_w = 3.76eE/F_y = 106.72$ (λ_p for compactness/noncompact) or 161.78 (λ_r for noncompact/slender). For a compact section $h/t_w = 106.72$, therefore $t_w = 68/106.72 = 0.637$ in. For a noncompact section web $t_w = 68/161.78 = 0.42$ in. say ½ in. Use a ½-in. (13-mm) plate. Hence, the area of the web $A_w = 34$ in.^{2} (219.35 cm^2).

3. Select flange plates using
From Table 4.1b of the AISC Specifications for compact flanges, $b/t \leq 0.38eE/F_y = 10.78$. Assuming a plate width of $b = 22$ in. (558.8 mm), flange thickness, t_f, required for compactness $t_f = 11/10.78 = 1.02$ in. Use $t_f = 1¼$-in. thick flange plate. Area of flange, $A_f = 22 \times 1.25 = 27.5$ in.^{2} (177.4 cm^2).

4. Check adequacy of trial section
Since section components (flanges and web) are compact, the next step is to compute the plastic section modulus, Z, of the section. $Z_{section} = Z_{flanges} + Z_{web} = 2A_f(h_{flange}/2) + 2[(h_{web}/2)(t_w)(h_{web}/4)] = 2 \times 27.5 \times (68 + 0.625 + 0.625)/2 + 2 \times [(68/2) \times (.5) \times (68/4)] = 2482.375$ in.^{3} (40,678.84 cm^3). Compute available moment capacity, $\phi M_n = \phi ZF_y = 0.9 \times 2482.375 \times 36 = 80,428.95$ kip-in. $= 6702.4$ kip · ft (9087.25 kN · m). The maximum moment of $1.6 \times 4053 = 6484.8$ kip · ft. Also, the available shear strength, ϕV_n, must equal or exceed the required shear strength, V_u. The nominal shear strength, V_n, of unstiffened or stiffened webs is given by $V_n = 0.6 \, F_y A_w C_v$. When $h/t_w \leq 1.10 \sqrt{(k_v E/F_y)}$, then $C_v = 1.0$.
Assume $k_v = 5$, maximum spacing. Hence $1.10 \times (5 \times 29000/36)^{0.5} = 69.8 < 136$. Also $1.37 \sqrt{(k_v E/F_y)} = 86.9 < 136$. Therefore, $C_v = \dfrac{1.51 \, k_v E}{(h/t_w)^2 F_y} = (1.51 \times 5 \times 29000/136^2 \times 36) = 0.33$.
Therefore $\phi V_n = 0.9(0.6 \times 36 \times 34 \times .33) = 218.12$ kips (970.62 kN) $< 1.6 \times 207$ kips $= 331.2$ kips (1473.84 kN). Section is adequate for bending but not shear. Stiffeners will be required to increase available shear strength.

5. Determine the distance of the stiffeners from the girder ends
Compute allowable shear stress per Table 3-16a of the AISC *Manual*, $\phi V_n/A_w = 218.12/34 = 6.4$ ksi. With stress value of 6.4 ksi and h/t_w of 136, $a/h = 1.35$. Therefore $a = (a/h) \times h = 1.35 \times 68 = 91.8$ in. $= 7$ ft 8 in. from girder ends.

6. Ascertain whether additional intermediate stiffeners are required
Transverse stiffeners are not required when $h/t_w \leq 2.46 \sqrt{(E/F_y)} = 2.46 \times (29000/36)^{0.5} = 69.82 < 136 \, (h/t_w)$. Therefore stiffeners are required. At E, the required shear strength, $V_u = 1.6 \times [207 - 4.67(4)] = 301.3$ kips (1339.3 kN). From the previous calculation $\phi V_n = 218.12$ kips (970.62 kN) < 301.3 kips (1339.3 kN), therefore intermediate stiffeners are required.

7. Provide stiffeners, and investigate the suitability of their tentative spacing
Provide stiffeners at F, the center of EB. See whether this spacing satisfies the Specification. Thus $[260/(h/t_w)]^2 = (260/136)^2 = 3.65$; $a/h = 122/68 = 1.79 < 3.65$. This is acceptable. Entering the table referred to in step 6 with $a/h = 1.79$ and $h/t_w = 136$ shows $V_{allow} = 10.5$ ksi > 6.32 ksi. This is acceptable.

8. Analyze the combination of shearing and bending stress
It has been shown that the interaction between shear and flexural resistance is negligible when the following equations are satisfied: $2A_w/(A_{fc} + A_{ft}) \leq 2.5$ and $h/b_f \leq 6$. $2 \times (68 \times .5)/(22 \times 1.25 + 22 \times 1.25) = 1.24 < 2.5$ and $68/22 = 3.09 < 6.0$, therefore no shear and moment interaction will be considered.

9. Investigate the need for transverse stiffeners in the center interval
Basically the available shear strength should be compared to the required shear strength for that panel, in this case BC. At B the shear force, $V = 32$ kips (142.3 kN). The required

shear strength, $V_u = 1.6 \times 32 = 52.14$ kips (227.38 kN). The available shear strength, $\phi V_n = 218.12$ kips. Therefore no intermediate stiffeners are required. Alternatively, the shear stress, v, can be computed from 32 kips/34 in.$^2 = 0.94$ ksi (6.48 MPa). $h/t_w = 68/.5 = 136$ in. (354.44 cm) and $a/h = 122/68 = 1.79$. Entering the allowable shear stress table for $F_y = 36$ ksi with tension field action give an allowable shear stress of over 10.5 ksi (72.4 MPa). The allowable shear then becomes $V = 10.5$ ksi $\times 34$ in.^{2} (A_w) = 357 kips > 32 kips.

10. Design the intermediate stiffeners in accordance with the Specification
Section G2.2 of the AISC Specification states that transverse intermediate stiffeners shall have a moment of inertia, I_{st}, about an axis in the web center for stiffener pairs or about the face in contact with the web plate for single stiffeners, which shall not be less than minimum $I_{st} = t_w^3 j$, where $j = [2.5/(a/h)^2] - 2 \geq 0.5$ [(AISC Eq. (G2-6). $a = 122$ in. (310 cm), $a/h = 1.79$, $t_w = 0.5$ in. (13 mm), $j = 2.5/[1.79^2] - 2 = -1.22$. Use $j = 0.5$. Minimum $I_{st} = 122 \times .53 \times (0.5) = 7.625$ in.^{4} (317.4 cm^4). Section G3 of the AISC Specification states that transverse stiffeners subject to tension field action must meet the following limitations: $(b/t)_{st} \leq 0.56 (E/F_{yst})^{1/2}$. AISC Eq. (G3-3) states that the area of the stiffeners, $A_{st} > (F_{yw}/F_{yst})[0.15 D_s h t_w (1 - C_v)(V_r/V_c) - 18_{tw}^2] \geq 0$, where $(b/t_w)_{st}$ = width-thickness ratio of the stiffener, F_{yst} = specified minimum yield stress of the stiffener material, C_v = coefficient defined in Section G2.1 of the AISC Specification $D_s = 1.0$ for stiffeners in pairs, 1.8 for single angle stiffeners and 2.4 for single plate stiffeners. V_r = required shear strength at the location of the stiffener and V_c = available shear strength ($\phi_v V_n$). Try a pair of ½ in. × 4 in. plates. $I_{st} = 2[(½)(4)^3/12 + (½)(4)[4/2 + (½)/2] = 14.33$ in.^{4} (596.6 cm^4). This is acceptable. The stiffeners must be in intimate contact with the compression flange, but they may terminate 1¾ in. (44.45 mm) from the compression flange. The connection of the stiffeners to the web must transmit the vertical shear specified in the *Specification*.

11. Design the bearing stiffeners at the supports
Try two plate stiffeners ¾ × 10 as shown in Fig. 9. Check width-thickness ratio of the stiffener plate. $b/t = 10.00/(¾) = 13.33 < 0.56(E/F_{yst})^{1/2} = 0.56(29,000/36)^{1/2} = 15.89$ OK. Check the compression strength of the bearing stiffener. Moment of Inertia, $I = 2[\frac{1}{12} \times (¾) \times (10^3)] = 125$ in.^{4} (5202 cm^4). Area of column $= 2(10)(¾) + 12 \times (0.5)(0.5) = 18$ in.^{2} (116.13 cm^2). Radius of gyration, $r = (I/A)^{1/2} = (125/18)^{1/2} = 2.64$ in. (6.7 cm). Effective

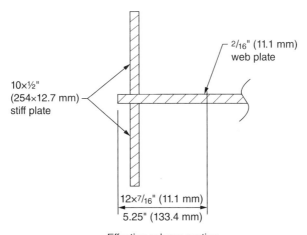

FIGURE 9. Effective column section.

length, $KL = 0.75h = 0.75(68) = 51$ in. (129.54 cm). $KL/r = 51/2.64 = 19.32$. From AISC Table 4-22 (for $KL/r = 19.32$ (say 19), $F_y = 36$ ksi): $< \phi_c F_{cr} = 31.8$ ksi (219.2 MPa). $< \phi_c P_n = < \phi_c F_{cr} A = 31.8(18) = 572.4$ kips $> R_u = 1.6 \times 207 = 331.2$ kips (1473.84 kN) OK. Check bearing criteria per AISC manual. $\phi P_n = (0.75)1.8 F_y A_{pd} = 0.75 \times 1.8 \times 36 \times (10 - \frac{1}{2})$ assuming $\frac{1}{2}$ in. is cut from the corners of the flange-to-web weld. $\phi P_n = 461.7$ kips (2054.6 kN) > 331.2 kips (OK). The $10 \times \frac{3}{4}$ in. (254.0)(19 mm) stiffeners at the supports are therefore satisfactory with respect to both column action and bearing.

Steel Columns and Tension Members

The general remarks appearing at the opening of the previous part apply to this part as well. A column is a compression member having a length that is very large in relation to its lateral dimensions. The *effective* length of a column is the distance between adjacent points of contraflexure in the buckled column or in the imaginary extension of the buckled column, as shown in Fig. 10. The column length is denoted by L, and the effective length by KL. Recommended design values of K are given in the AISC *Manual*.

The capacity of a column is a function of its effective length and the properties of its cross section. It therefore becomes necessary to formulate certain principles pertaining to the properties of an area. Consider that the moment of inertia I of an area is evaluated with respect to a group of concurrent axes.

There is a distinct value of I associated with each axis, as given by earlier equations in this section. The *major* axis is the one for which I is maximum; the *minor* axis is the one for which I is minimum. The major and minor axes are referred to collectively as the *principal* axes. With reference to the equation given earlier, namely, $I_{x''} = I_{x'} \cos^2\theta + I_{y'} \sin^2\theta - P_{x''y''} \sin 2\theta$, the orientation of the principal axes relative to the given x' and y' axes is found by differentiating $I_{x''}$ with respect to θ, equating this derivative to zero, and solving for θ to obtain $\tan 2\theta = 2P_{x''y''}/(I_{y'} - I_{x'})$, Fig. 15. The following statements are corollaries of this equation:

1. The principal axes through a given point are mutually perpendicular, since the two values of θ that satisfy this equation differ by $90°$.

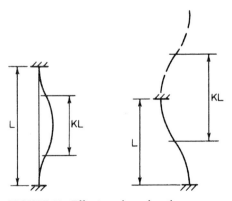

FIGURE 10. Effective column lengths.

2. The product of inertia of an area with respect to its principal axes is zero.

3. Conversely, if the product of inertia of an area with respect to two mutually perpendicular axes is zero, these are principal axes.

4. An axis of symmetry is a principal axis, for the product of inertia of the area with respect to this axis and one perpendicular thereto is zero.

Let A_1 and A_2 denote two areas, both of which have a radius of gyration r with respect to a given axis. The radius of gyration of their composite area is found in this manner: $I_c = I_1 + I_2 = A_1 r^2 + A_2 r^2 = (A_1 + A_2) r^2$. But $A_1 + A_2 = A_c$. Substituting gives $I_c = A_w r^2$; therefore, $r_c = r$. This result illustrates the following principle: If the radii of gyration of several areas with respect to a given axis are all equal, the radius of gyration of their composite area equals that of the individual areas. The equation $I_x = \Sigma I_0 + \Sigma A k^2$, when applied to a single area, becomes $I_x = I_0 + Ak^2$. Then $Ar_x^2 = Ar_o^2 + Ak^2$, or $r_x = (r^2 + k^2)^{0.5}$. If the radius of gyration with respect to a centroidal axis is known, the radius of gyration with respect to an axis parallel thereto may be readily evaluated by applying this relationship. The Euler equation for the strength of a slender column reveals that the member tends to buckle about the minor centroidal axis of its cross section. Consequently, all column design equations, both those for slender members and those for intermediate-length members, relate the capacity of the column to its minimum radius of gyration. The first step in the investigation of a column, therefore, consists in identifying the minor centroidal axis and evaluating the corresponding radius of gyration.

CAPACITY OF A BUILT-UP COLUMN

A compression member consists of two C15 × 40 channels laced together and spaced 10 in. (254.0 mm) back to back with flanges outstanding, as shown in Fig. 11. What axial load may this member carry if its effective length is 22 ft (6.7 m)?

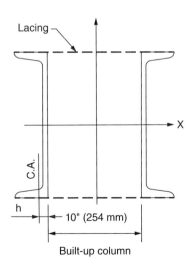

FIGURE 11. Built-up column section.

Calculation Procedure:

1. Record the properties of the individual channel
Since x and y are axes of symmetry, they are the principal centroidal axes. However, it is not readily apparent which of these is the minor axis, and so it is necessary to calculate both r_x and r_y. The symbol r, without a subscript, is used to denote the minimum radius of gyration, in inches (centimeters). Using the AISC *Manual*, we see that the channel properties are $A = 11.80$ in.^{2} (76.129 cm^2); $h = 0.77$ in. (19.558 mm); $r_1 = 5.43$ in. (137.922 mm); $r_2 = 0.88$ in. (22.352 mm).

2. Evaluate the minimum radius of gyration of the built-up section
Determine the slenderness ratio. Thus, $r_x = 5.43$ in. (137.922 mm); $r_y = (r_2^2 + 5.77^2)^{0.5} > 5.77$ in. (146.558 mm); therefore, $r = 5.43$ in. (137.922 mm); $KL/r = 22(12)/5.43 = 48.6$. Assuming $K = 1$, and $k = 10/2 + h = 5 + 0.77 = 5.77$ in. (146.558 mm).

3. Determine the available critical stress in the column
Enter the AISC *Manual* Table 4-22. Available Critical Stress for Compression Members with a slenderness ratio of 48.6 (say 49 – conservative) to obtain the critical stress $\phi F_{cr} = 28.5$ kips/in^2 (196.5 MPa). Then, the column capacity $= P = A \times \phi F_{cr} = 2(11.80)(28.5) = 672.6$ kips (2986 kN).

CAPACITY OF A DOUBLE-ANGLE STAR STRUT

A star strut is composed of two $5 \times 5 \times \frac{3}{8}$ in. (127.0 × 127.0 × 9.53 mm) angles intermittently connected by $\frac{3}{8}$-in. (9.53-mm) batten plates in both directions. Determine the capacity of the member for an effective length of 12 ft (3.7 m).

Calculation Procedure:

1. Identify tile minor axis
Refer to Fig. 12a. Since p and q are axes of symmetry, they are the principal axes; p is manifestly the minor axis because the area lies closer to p than q.

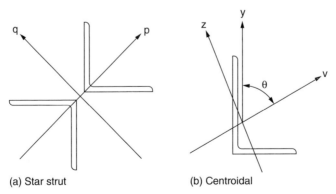

(a) Star strut   (b) Centroidal

FIGURE 12. (*a*) Star strut; (*b*) centroidal axes of angle section.

2. Determine r_v

Refer to Fig. 12b, where v is the major and z the minor axis of the angle section. Apply $I_{x''} = I_{x'} \cos^2 \theta + I_{y'} \sin^2 \theta - P_{x'y'} \sin^2 \theta$, and set $P_{vz} = 0$ to get $r_y^2 = r_v^2 \cos^2 \theta + r_x^2 \sin^2 \theta$, therefore, $r_v^2 \sec^2 \theta - r_x^2 \tan^2 \theta$. For an equal-leg angle, $\theta = 45°$, and this equation reduces to $r_v^2 = 2r_y^2 - r_z^2$.

3. Record the member area and computer r_v

From the *Manual*, $A = 3.61$ in.^{2} (23.291 cm^2); $r_y = 1.56$ in. (39.624 mm); $r_z = 0.99$ in. (25.146 mm); $r_v = (2 \times 1.56^2 - 0.99^2)^{0.5} = 1.97$ in. (50.038 mm).

4. Determine tile minimum radius of gyration of tile built-up section; compute the strut capacity

Thus, $r = r_p = 1.97$ in. (50.038 mm); $KL/r = 12(12)/1.97 = 73$. From the AISC *Manual* Table 4-22, $\phi F_{cr} = 24.5$ kips/in^2 (168.922 MPa). Then $P = A\phi F_{cr} = 2(3.61)(24.5) = 176.89$ kips (786.846 kN).

SECTION SELECTION FOR A COLUMN WITH TWO EFFECTIVE LENGTHS

A 30-ft-long (9.2-m) column is to carry a 200-kip (889.6-kN) live load. The column will be braced about both principal axes at top and bottom and braced about its minor axis at midheight. Architectural details restrict the member to a nominal depth of 8 in. (203.2 mm). Select an A242 steel section with yield strength of 50 ksi, by consulting the available strength in axial compression tables in the AISC LRFD *Manual* and then verify the design.

Calculation Procedure:

1. Compute the required strength
$P_u = 1.6 \times 200 = 320$ kips (1423 kN).

2. Select a column section
Refer to Fig. 13. The effective length with respect to the minor axis may be taken as 15 ft (4.6 m). Then $K_x L = 30$ ft (9.2 m) and $K_y L = 15$ ft (4.6 m). The available strengths in axial compression recorded in the *Manual* tables are calculated on the premise that the column tends to buckle about the minor axis. In the present instance, however, this premise is not necessarily valid. It is expedient for design purposes to conceive of a uniform-strength column, i.e., one for which K_x and K_y bear the same ratio as r_x and r_y, thereby endowing the column with an identical slenderness ratio with respect to the two principal axes. Select a

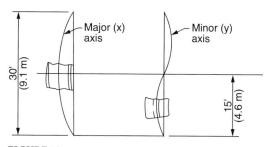

FIGURE 13

column section on the basis of the K_yL value; record the value of r_x/r_y of this section. Table 4-1 of the AISC *Manual* shows that a W8 × 48 column has available strength in axial compression of 367 kips (1632 kN) when $K_yL = 15.0$ ft (4.6 m); at the bottom of the table it is found that $r_x/r_y = 1.74$.

3. Compute the equivalent y-y axis effective length for strong axis buckling

Thus, $K_xL = 30.0/1.74 = 17.34$ ft (5.35 m). The available strength from Table 4-1 of the *Manual* for a W4 × 48 with an effective length of 17 ft = 314 kips (1397 kN) < 320 kips (1423 kN) required strength. Therefore section is not adequate and capacity is governed by the equivalent *y-y* strong axis buckling.

4. Try a specific column section of larger size

Trying W8 × 58, the available strength for 17 ft effective length = 386 kips (1717 kN) > 320 kips, therefore section is adequate.

5. Verify the design

To verify the design, record the properties of this section W8 × 58 and compute the slenderness ratios. For this grade of steel and thickness of member, the yield-point stress is 50 kips/in.² (344.8 MPa), as given in the *Manual*. Thus, $A = 17.1$ in.² (110 cm²); $r_x = 3.65$ in. (92.8 mm); $r_y = 2.10$ in. (53.34 mm). Then $K_xL/r_x\ 30(12)/3.65 = 98.63$; $K_yL/r_y = 15(12)/2.10 = 85.7$.

6. Determine the available critical stress, ϕF_{cr}, and member capacity

From the *Manual*, $\phi F_{cr} = 22$ kips/in.² (151.685 MPa) with a slenderness ratio of 99. Then $P = 17.1(22) = 376.2$ kips (1673 kN). Therefore, use W8 × 58 because the capacity of the column exceeds the intended load.

STRESS IN COLUMN WITH PARTIAL RESTRAINT AGAINST ROTATION

The beams shown in Fig. 14a are rigidly connected to a W14 × 95 column of 28 ft (8.5 m) height that is pinned at its foundation. The column is held at its upper end by cross bracing lying in a plane normal to the web. Compute the available critical stress in the column in the absence of bending stress.

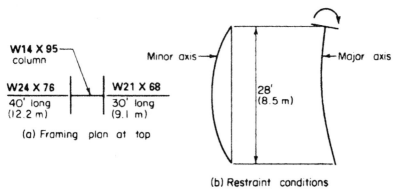

FIGURE 14

Calculation Procedure:

1. Draw schematic diagrams to indicate the restraint conditions
Show these conditions in Fig. 14b. The cross bracing prevents sidesway at the top solely with respect to the minor axis, and the rigid beam-to-column connections afford partial fixity with respect to the major axis.

2. Record the I_x values of the column and beams

Section	I_x	
	in.^{4}	cm^4
W14 × 95	1064	44,287
W24 × 76	2096	87,242
W21 × 68	1478	61,519

3. Calculate the rigidity of the column relative to that of the restraining members at top and bottom
Thus, $I_c/L_c = 1064/28 = 38$. At the top, $\Sigma(I_g/L_g)$ 2096/40 + 1478/30 = 101.7. At the top, the rigidity $G_t = 38/101.7 = 0.37$. In accordance with the Commentary 7.2 of the Specifications in the *Manual*, under "Adjustments for Columns with Differing End Conditions," for column ends supported by, but not rigidly connected to, a footing or foundation, G is theoretically infinity but unless designed as a true friction-free pin, may be taken as 10 for practical designs. Therefore set the rigidity at the bottom $G_b = 10$.

4. Determine the value of K_x
Using the *Manual* alignment chart, Fig. C-A-7.2, determine that $K_x = 1.77$.

5. Compute the slenderness ratio with respect to both principal axes, and find the available critical stress ϕF_{cr}
Thus, $K_x L/r_x = 1.77(28)(12)/6.17 = 96.4$; $K_y L/r_y = 28(12)/3.71 = 90.6$. Using the larger value of the slenderness ratio, find from the *Manual* the available critical stress $\phi F_{cr} = 22.9$ ksi (157.89 MPa).

LACING OF BUILT-UP COLUMN

Design the lacing bars and end tie plates of the member in Fig. 15. The lacing bars will be connected to the channel flanges with ½-in. (12.7-mm) bolts.

Calculation Procedure:

1. Establish the dimensions of the lacing system to conform to the AISC Specification
The function of the lacing bars and tie plates is to preserve the integrity of the column and to prevent local failure. Refer to Fig. 15. The standard gage in 15-in. (381.0-mm) channel = 2 in. (50.8 mm), from the AISC *Manual*. Then $h = 14 < 15$ in. (381.0 mm); therefore, use single lacing. Try $\theta = 60°$; then, $v = 2(14) \cot 60° = 16.16$ in. (410.5 mm). Set $v = 16$ in. (406.4 mm); therefore, $d = 16.1$ in. (408.94 mm). For the built-up section, $KL/r = 48.5$; for the single channel, $KL/r = 16/0.89 < 48.5$. This is acceptable. The spacing of the bars is therefore satisfactory.

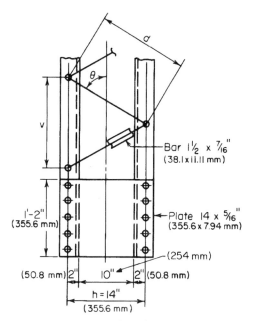

FIGURE 15. Lacing and tic plates.

2. Design the lacing bars
The lacing system must be capable of transmitting an assumed transverse shear equal to 2 percent of the available compressive strength of the member; this shear is carried by two bars, one on each side. A lacing bar is classified as a secondary member. To compute the transverse shear, assume that the column will be loaded to its capacity of 432 kips (1921.5 kN). Then force per bar = ½ (0.02)(432) × (16.1/14) = 5.0 kips (22.24 kN). Also, $L/r <$ 140; therefore, $r = 16.1/140 = 0.115$ in. (2.9210 mm). For a rectangular section of thickness t, $r = 0.289t$ [i.e. $\sqrt{(bt^3/12/(bt))}$]. Then $t = 0.115/0.289 = 0.40$ in. (10.160 mm). Set $t = 7/16$ in. (11.11 mm); $r = 0.127$ in. (3.226 mm); $L/r = 16.1/0.127 = 127$; $\phi F_{cr} = 14$ ksi (96.52 MPa); $A = 5.0/14 = 0.36$ in.^{2} (2.30 cm^2). From the *Manual*, the minimum width required for ½-in. (12.7-mm) bolts 1½ in. (38.1 mm). Therefore, use a flat bar 1½ × 7/16 in. (38.1 × 11.11 mm); $A = 0.66$ in.^{2} (4.258 cm^2).

3. Design the end tie plates in accordance with the Specification
The minimum length 14 in. (355.6 mm); $t = 14/50$ (AISC Manual Part 16 Section E6.2) = 0.28. Therefore, use plates 14 × 5/16 in. (355.6 × 7.94 mm). The bolt pitch is limited to six diameters, or 3 in. (76.2 mm).

SELECTION OF A COLUMN WITH A LOAD AT AN INTERMEDIATE LEVEL

A column of 30-ft (9.2-m) length carries a load of 130 kips (578.2 kN) applied at the top and a load of 56 kips (249.1 kN) applied to the web at mid-height. Select an 8-in. (203.2-mm) column of A242 steel, using $K_x L = 30$ ft (9.2 m) and $K_y L = 15$ ft (4.6 m).

Calculation Procedure:

1. Compute the effective length of the column with respect to the major axis
The following procedure affords a rational method of designing a column subjected to a load applied at the top and another load applied approximately at the center. Let m = load at intermediate level, kips per total load, kips (kilonewtons). Replace the factor K with a factor K' defined by $K' = K(1 - m/2)^5$. Thus, 0.5 for this column, $m = 56/186 = 0.30$. And $K'_x L = 30(1-0.15)^{0.5} = 27.6$ ft (8.41 m).

2. Select a trial section on the basis of the $K_y L$ value
From the AISC *Manual* for a W8 × 40, capacity = 186 kips (827.3 kN) when $K_y L = 16.2$ ft (4.94 m) and $r_x/r_y = 1.73$.

3. Determine whether the selected section is acceptable
Compute the value of $K_x L$ associated with a uniform-strength column, and compare this with the actual effective length. Thus, $K_x L = 1.73(16.2) = 28.0 > 27.6$ ft (8.41 m). Therefore, the W8 × 40 is acceptable.

DESIGN OF AN AXIAL MEMBER FOR FATIGUE

A web member in a welded truss will sustain precipitous fluctuations of stress caused by moving loads. The structure will carry three load systems having the following characteristics: Force induced in member, kips (kN).

System	Max Compression	Maximum Tension	No. of Times Applied
A	46 (204.6)	18 (80.1)	60,000
B	40 (177.9)	9 (40.0)	1,000,000
C	32 (142.3)	8 (35.6)	2,500,000

The effective length of the member is 11 ft (3.4 m). Design a double-angle member.

Calculation Procedure:

1. Calculate for each system the design load, and indicate the yield-point stress on which the allowable stress is based
The design of members subjected to a repeated variation of stress is regulated by the AISC *Specification*. For each system, calculate the design load and indicate the yield-point stress on which the allowable stress is based. Where the allowable stress is less than that normally permitted, increase the design load proportionately to compensate for this reduction. Let (+) denote tension and (−) denote compression. Then

System	Design Load, kips (KN)	Yield-Point Stress, ksi (MPa)
A	−46 − ⅔(18) = −58 (−257.9)	36 (248.2)
B	−40 − ⅔(9) = −46 (−204.6)	33 (227.5)
C	1.5(−32 − ¾ × 8) = −57 (−253.5)	33 (227.5)

2. Select a member for system A and determine if it is adequate for system C

From the AISC *Manual*, try two angles 4 × 3½ × ⅜ in. (101.6 × 88.90 × 9.53 mm), with long legs back to back; the capacity is 65 kips (289.1 kN). Then $A = 5.34$ in.² (34.453 cm²); $r = r_x = 1.25$ in. (31.750 mm); $KL/r = 11(12)/1.25 = 105.6$. From the *Manual*, for a yield-point stress of 33 ksi (227.5 MPa), $\sigma = 11.76$ ksi (81.085 MPa). Then the capacity $P = 5.34(11.76) = 62.8$ kips (279.3 kN) > 57 kips (253.5 kN). This is acceptable. Therefore, use two angles 4 × 3½ × ⅜ in. (101.6 × 88.90 × 9.53 mm), long legs back to back.

INVESTIGATION OF A BEAM COLUMN

A W12 × 53 column with an effective length of 20 ft (6.1 m) is to carry an axial live load of 160 kips (711.7 kN) and live load moments indicated in Fig. 16. The member will be secured against side-sway in both directions. Is the section adequate? Assume $KL_x = KL_y = L_b = 20$ ft and yield stress, $F_y = 50$ ksi.

Calculation Procedure:

1. Compute required strengths
Required axial strength, $P_u = 1.6 \times 160$ kips $= 256$ kips (1139 kN). Assume end moments are about the strong axis, $M_{ux} = 1.6 \times 31.5$ kip · ft $= 50.4$ kip · ft (68.33 kN · m) and $M_{uy} = 1.6 \times 15.2$ kip · ft $= 24.32$ kip · ft (32.97 kN · m).

2. Record the properties of the section
The properties of the section from Table 1-1 of the AISC *Manual* are $A = 15.59$ in.² (100.586 cm²); $S = 70.7$ in.³ (1158.77 cm³); $r_x = 5.23$ in. (132.842 mm); $r_y = 2.48$ in. (62.992 mm).

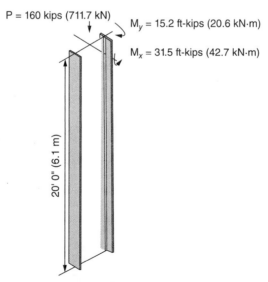

FIGURE 16. Beam column.

3. Record available axial and flexural strengths
From Tables 4-1, 3-10, and 3-4 of the AISC *Manual* at $KL_y = 20$ ft, $P_c = \phi_c P_n = 354$ kips (1,575 kN). At $L_b = 20$ ft, $M_{cx} = \phi M_{nx} = 234$ kip · ft (317.261 kN · m). For a W12 × 53, $\phi_b M_{ny} = 109$ kip · ft and $\phi_b M_{nx} = 185$ kip · ft at L_r (unbraced length) = 28.2 ft.

4. Compute $P_r/P_c = 256/354 = 0.723 > 0.2$ therefore use equation H1-1a as follows:

$$\frac{P_r}{P_c} + \frac{8}{9}\left(\frac{M_{rx}}{M_{cx}} + \frac{M_{ry}}{M_{cy}}\right) \leq 1.0$$

$$= 0.723 + 8/9[(50.4/185) + (24.32/109)] = 1.16 \text{ say } 1.0.$$

This section is therefore satisfactory.

APPLICATION OF BEAM-COLUMN FACTORS USING ASD METHOD

For the previous calculation procedure, investigate the adequacy of the W12 × 53 section by applying the values of the beam-column factors B and a given in the AISC *Manual*.

Calculation Procedure:

1. Record the basic values of the previous calculation procedure
The beam-column factors were devised in an effort to reduce the labor involved in analyzing a given member as a beam column when $f_a/F_a > 0.15$. They are defined by $B = A/S$ per inch (decimeter); $a = 0.149 \times 10^6$ (Ar^2) in.^{4} (6201.9I dm^4). Let P denote the applied axial load and P_{allow} the axial load that would be permitted in the absence of bending. The equations given in the previous procedure may be transformed to $P + [B_x M_x C_{mx}(F_a/F_{bx}) \times (a_x/[a_x - P(KL)^2]] + [B_y M_y C_{my}(F_a/F_{by}) \times (a_y/a_y - P(KL)^2]] \leq P$, and $P \times (F_a/(0.6F_y) + [B_x M_x (F_a/F_{bx}] + [B_y M_y (F_a/F_{by}] \leq P$, where KL, B, and a are evaluated with respect to the plane of bending. The basic values of the previous procedure are $P = 160$ kips (711.7 kN); $M = 31.5$ ft · kips (42.712 kN · m); $F_b = 22$ kips/in. (151.7 MPa); $C_m = 0.793$.

2. Obtain the properties of the section
From the *Manual* for a W12 × 53, $A = 15.59$ in.^{2} (100.587 cm^2); $B = 0.221$ per inch (8.70 per meter); $a = 63.5 \times 10^6$ in.^{4} (264.31 × 10^3 dm^4). Then when $KL = 20$ ft (6.1 m), $P_{\text{allow}} = 209$ kips (929.6 kN).

3. Substitute in the first transformed equation
Thus, $F = P/A = 209/15.59 = 13.41$ kips/in.^{2} (92.461 MPa), $P(KL)^2 = 160(240)^2 = 9.22 \times 10^6$ kip·in.^{2} (2.648 × 10^4 kN · m^2), and $a/[a - P(KL)^2] = 63.5/(63.5 - 9.22) = 1.17$.

Then $160 + 0.221(31.5)(12)(0.793)(13.41/22)(1.17) = 207 < 209$ kip (929.6 kN). This is acceptable.

4. Substitute in the second transformed equation
Thus, $160(13.41/22) + 0.221(31.5)(12)(13.41/22) = 148 < 209$ kips (929.6 kN). This is acceptable. The W12 × 53 section is therefore satisfactory.

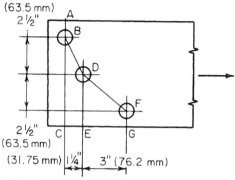

FIGURE 17

NET SECTION OF A TENSION MEMBER

The 7 × ¼ in. (177.8 × 6.35 mm) plate in Fig. 17 carries a tensile force of 18,000 lb (80,064.0 N) and is connected to its support with three ¾-in. (19.05-mm) rivets in the manner shown. Compute the maximum tensile stress in the member.

Calculation Procedure:

1. Compute the net width of the member at each section of potential rupture

The AISC *Specification* prescribes the manner of calculating the net section of a tension member. The effective diameter of the holes is considered to be ⅛ in. (3.18 mm) greater than that of the rivets. After computing the net width of each section, select the minimum value as the effective width. The *Specification* imposes an upper limit of 85 percent of the gross width.

Refer to Fig. 17. From B to D, $s = 1.25$ in. (31.750 mm), $g = 2.5$ in. (63.50 mm); from D to F, $s = 3$ in. (76.2 mm), $g = 2.5$ in. (63.50 mm); $w_{AC} = 7 - 0.875 = 6.12$ in. (155.45 mm); $W_{ABDE} = 7 - 2(0.875) + 1.25^2/[4(2.5)] = 5.41$ in. (137.414 mm); $w = 7 - 3(0.875) + 1.25^2/(4 \times 2.5) + 3^2/(4 \times 2.5) = 5.43$ in. (137.922 mm); $w_{max} = 0.85(7) = 5.95$ in. (151.13 mm). Selecting the lowest value gives $w_{eff} = 5.41$ in. (137.414 mm).

2. Compute the tensile stress on the effective net section

Thus $f = 18,000/[5.41(0.25)] = 13,300$ lb/in.^{2} (91,703.5 kPa).

DESIGN OF A DOUBLE-ANGLE TENSION MEMBER

The bottom chord of a roof truss sustains a tensile force of 141 kps (627.2 kN). The member will be spliced with ¾-in. (19.05-mm) rivets as shown in Fig. 18a. Design a double-angle member and specify the minimum rivet pitch.

Calculation Procedure:

1. Show one angle in its developed form

Cut the outstanding leg, and position it to be coplanar with the other one, as in Fig. 18b. The gross width of the angle w_g is the width of the equivalent plate thus formed; it equals the sum of the legs of the angle less the thickness.

2. Determine the gross width in terms of the thickness

Assume tentatively that 2.5 rivet holes will be deducted to arrive at the net width. Express w_g in terms of the thickness t, of each angle. Then net area required = 141/22 = 6.40 in.² (41.292 cm²); also, $2t(w - 2.5 \times 0.875)$ 6.40; $w_g = 3.20/t + 2.19$.

3. Assign trial thickness values, and determine the gross width

Construct a table of the computed values. Then select the most economical size of member. Thus the most economical member is the one with the least area. Therefore, use two angles 6 × 4 × 7/16 in. (152.4 × 101.6 × 11.11 mm).

t, in. (mm)	w_g, in. (mm)	$w_g + t$, in. (mm)	Available size, in. (mm)	Area, in.² (cm²)
½ (12.7)	8.59 (218.186)	9.09 (230.886)	6 × 3½ × ½ (152.4 × 88.9 × 12.7)	4.50 (29.034)
7/16 (11.11)	9.50 (241.300)	9.94 (252.476)	6 × 4 × 7/16 (152.4 × 101.6 × 11.11)	4.18 (26.969)
⅜ (9.53)	10.72 (272.228)	11.10 (281.940)	None	

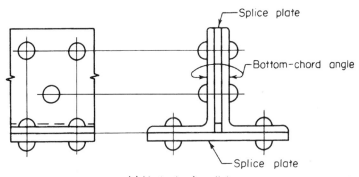

(a) Method of splicing

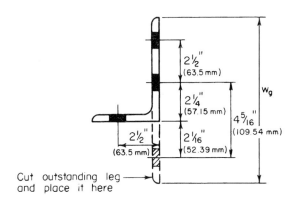

(b) Development of angle for net section

FIGURE 18

1.38   STRUCTURAL ENGINEERING

4. Record the standard gages
Refer to the *Manual* for the standard gages, and record the values shown in Fig. 18*b*.

5. Establish the rivet pitch
Find the minimum value of s to establish the rivet pitch. Thus, net width required = $\frac{1}{2}[6.40/(7/16)] = 7.31$ in. (185.674 mm); gross width = $6 + 4 - 0.44 = 9.56$ in. (242.824 mm). Then $9.56 - 3(0.875) + s^2/(4 \times 2.5) + s^2/(4 \times 4.31) = 7.31$; $s = 1.55$ in. (39.370 mm).

For convenience, use the standard pitch of 3 in. (76.2 mm). This results in a net width of 7.29 in. (185.166 mm); the deficiency is negligible.

Load and Resistance Factor Method

Abraham J. Rokach, MSCE, Associate Director of Education, American Institute of Steel Construction, Inc., writing in *Theory and Problems of Structural Steel Design*, McGraw-Hill, states "In 1986 a new method of structural steel design was introduced in the United States with the publication of the *Load and Resistance Factor Design Specification for Structural Steel Buildings*. Load and resistance factor design, (LRFD) has joined the old allowable stress design (ASD) method as a recognized means for the design of structural steel frameworks for buildings.

"Although ASD has enjoyed a long history of successful usage and is familiar to engineers and architects, the author and most experts prefer LRFD because it is a truer representation of the actual behavior of structural steel and unlike ASD, it can provide equivalent margins of safety for all structures under all loading condition. . . . For these reasons its anticipated that LRFD will replace ASD as the standard method of structural steel design."

The following selected procedures in this handbook cover structural steel design for buildings using the LRFD method drawn from the excellent Rokach book listed above. And competent authorities on the LRFD method, listed below, are cited frequently in the Rokach book, and in this handbook, usually in abbreviated form:

AISC: American Institute of Steel Construction, Inc., Chicago, IL.

AISC *LRFD Specification: Load and Resistance Factor Design Specification for Structural Steel Buildings*, published by AISC.

AISC *LRFD Manual: Load and Resistance Factor Design Manual of Steel Construction*, also published by AISC.

Equations in the following calculation procedures in this handbook are numbered as follows. Those equations appearing in the AISC LRFD Specification are accompanied by their AISC numbers in parentheses, thus ( ); other equations are numbered in brackets, thus [ ].

It is recommended that the designer have copies of both the AISC *LRFD Specification* and the AISC *Manual* on his or her desk when preparing any structural steel design using the LRFD method. Both are available from the AISC at 1 E Wacker Dr, Suite 3100, Chicago IL 60601. Abraham J. Rokach writes, further in his book cited above, "The ASD method is characterized by the use of one judgemental factor of safety. A limiting stress (usually F_y) is divided by a factor of safety (FS, determined by the authors of the *Specification*) to arrive at an allowable stress.

$$\text{Allowable stress} = F_y/\text{FS}$$

Actual stresses in a steel member are calculated by dividing forces or moments by the appropriate section property (e.g., area or section modulus). The actual stresses are then compared with the allowable stresses to ascertain that

$$\text{Actual stress} = \text{allowable stress}$$

No distinction is made among the various kinds of loads. Because of the greater variability and uncertainty of the live load and other loads in comparison with the dead load, a uniform reliability for all structures is not possible.

" ... Briefly, LRFD uses a different factor for each type of load and another factor for the strength or resistance. Each factor is the result of a statistical study of the variability of the subject quantity. Because the different factors reflect the degrees of uncertainty in the various loads and the resistance, a uniform reliability is possible."

DETERMINING IF A GIVEN BEAM IS COMPACT OR NONCOMPACT

A designer plans to use a W6 × 15 and a W12 × 65 beam in (a) A6 steel [F_y = 36 ksi (248 MPa)] and (b) with F_y = 50 ksi (344.5 MPa) and wishes to determine if the beams are compact or noncompact.

Calculation Procedure:

For the W6 × 15 Beam

1. Analyze the W6 × 15 beam

Refer to the AISC *Manual* Table B4. 1b, namely "Limiting Width-Thickness Ratios for Compression Elements Subject to Flexure" and its illustration "Definition of widths (b and h) and thickness," the flanges of a W shape are given by

$$\lambda_p = 0.38\sqrt{\frac{E}{F_y}} \text{ for compact section}$$

and

$$\lambda_r = 1.0\sqrt{\frac{E}{F_y}} \text{ for slender noncompact sections}$$

where λ_p = limiting width-thickness ratio for compact section and λ_r = for non-compact sections. Substituting for each of the two beams above we have

$$\lambda_p = 0.38\sqrt{\frac{29000 \text{ ksi}}{36 \text{ ksi}}} = 10.8 \quad \text{and} \quad \lambda_p = 0.38\sqrt{\frac{29000 \text{ ksi}}{50 \text{ ksi}}} = 9.2$$

2. Compute the data for the web of a W shape

Using the same equation as in step 1, for the web of a W shape

$$\lambda_p = 3.76\sqrt{\frac{E}{F_y}} \text{ for compact sections}$$

and

$$\lambda_r = 5.70\sqrt{\frac{E}{F_y}} \text{ for slender noncompact sections}$$

$$\lambda_p = 3.76\sqrt{\frac{29000 \text{ ksi}}{36 \text{ ksi}}} = 106.7 \quad \text{and} \quad \lambda_p = 5.70\sqrt{\frac{29000 \text{ ksi}}{50 \text{ ksi}}} = 90.5$$

3. Determine if the beam is compact
From the properties tables for W shapes, in Part 1 of the AISC *LRFD Manual* (Compact Section Criteria): for a W6 × 15, flange $b/2t_f = 5.99/2 \times .26) = 11.5$ and for the web $h_c/t_w = 21.6$. Since flange $(b/t = 11.5) > (\lambda_p = 10.8)$, the W × 15 beam is noncompact in A36 steel. Likewise, it is non-compact if $F_y = 50$ ksi (344.4 MPa).

For the W12 × 65 Beam

4. Compute the properties of the beam shape
From the AISC *Manual* "Properties Tables for W Shapes," for a W12 × 65, flange $b/2t_f = 9.92$ and for the web $h_c/t_w = 24.9$.

  a. In A36 steel.
  b. Since flange $(b/2t_f = 9.92) < (\lambda_p = 10.8)$ and web $(h_c/t_w = 24.9) < (\lambda_p = 106.7)$, a W12 × 65 beam is compact in A36 steel.
  c. However, if $F_y = 50$ ksi (344.5 MPa).

Because flange $(b/2t_f = 9.92) > (\lambda_p = 9.2)$, a W12 × 65 beam is non-compact if $F_y = 50$ ksi (344.5 MPa),

$$\text{Flange } \lambda_p = 9.2$$

$$\text{Web } \lambda_p = 90.5 \text{ (see W6} \times 15)$$

Because flange $(b/2t_f = 9.92) > (\lambda_p = 9.2)$, a W12 × 65 beam is noncompact if $F_y = 50$ ksi (344.5 MPa).

Related Calculations. The concept of compactness, states Abraham J. Rokach, MSCE, AISC, relates to local buckling. Cross sections of structural members are classified as compact, noncompact, or slender-element sections. A section is compact if the flanges are continuously connected to the web, and the width-thickness ratios of all its compression elements are equal to, or less than, λ_p. Structural steel members with compact sections can develop their full strength without local instability. In design, the limit state of local buckling need not be considered for compact members. This procedure is the work of Abraham J. Rokach, MSCE, AISC, associate director of education, American Institute of Steel Construction. SI values were prepared by the handbook editor.

DETERMINING COLUMN AXIAL SHORTENING WITH A SPECIFIED LOAD

A W10 × 49 column, 10 ft (3 m) long, carries a service load of 250 kips (113.5 mg). What axial shortening will occur in this column with this load?

Calculation Procedure:

1. Choose a suitable axial displacement equation for this column
The LRFD equation for axial shortening of a loaded column is

$$\text{Shortening } \Delta = \frac{Pl}{EA_g}$$

where Δ = axial shortening, in. (cm); P = unfactored axial force in member, kips (kg); l = length of member, in. (cm); E = modulus of elasticity of steel = 29,000 ksi (199.8 MPa); A_g = cross sectional area of member, sq. in. (sq. cm).

2. Compute the column axial shortening

Substituting,

$$\text{Shortening, } \Delta = \frac{Pl}{EA_g} = \frac{250 \text{ kips} \times (10 \text{ ft} \times 12 \text{ in./ft})}{29,000 \text{ kips/in.}^2 \times 14.4 \text{ in.}^2} = 0.072 \text{ in. } (0.183 \text{ cm})$$

Related Calculations. Use this equation to compute axial shortening of any steel column in LRFD work. This procedure is the work of Abraham J. Rokach, MSCE, American Institute of Steel Construction.

DETERMINING THE COMPRESSIVE STRENGTH OF A WELDED SECTION

The structural section in Fig. 19a is used as a 40-ft (12.2-m) column. Its effective length factor $K_x = K_y = 1.0$. Determine the design compressive strength if the steel is A36.

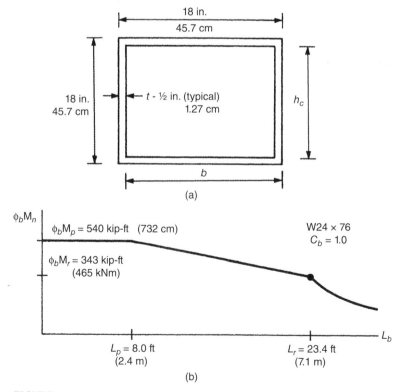

FIGURE 19

Calculation Procedure:

1. Choose a design compressive strength
The design available compressive strength is given by

$$\phi_c P_n = \phi_c F_{cr} A_g$$

The values of $\phi_c F_{cr}$ can be obtained from Table 4-22, "Available Critical Stress for Compression Members of 36 ksi Specified Yield-Stress Steel, $\phi_c = 0.85$" in the AISC *Manual*, if Kl/r is known. With $Kl = 1.0 \times 40.0$ ft $\times$ 12 in./ft = 480 in. (1219 cm), then $r = \sqrt{I/A}$, $A = (18 \text{ in.})^2 - (17 \text{ in.})^2 = 35 \text{ in.}^2$, $I_x = I_y = I = [(18 \text{ in.})^4 - (17 \text{ in.})^4]/12 = 1788 \text{ in.}^4$ (42.95 cm^4).

2. Find Kl/r ratio for this section
With the data we have

$$r = \sqrt{\frac{1788 \text{ in.}^4}{35 \text{ in.}^4}} = 7.15 \text{ in.}$$

$$\frac{Kl}{r} = \frac{480 \text{ in.}}{7.15 \text{ in.}} = 67.13$$

3. Determine the available design compressive strength of this section
Using the suitable AISC *Manual* table, namely "Available Critical Stress for Compression Members of 36 ksi Specified Yield-Stress Steel, $\phi_c = 0.85$." and interpolating, for $Kl/r = 67.2$, $\phi_c F_{cr} = 24.13$ ksi (166.3 MPa) the design compressive strength $\phi_c P_n = 24.13$ kips/in.$^2 \times$ 35.0 in.^{2} = 845 kips (3759 kN).

Related Calculations. This procedure is the work of Abraham J. Rokach, MSCE, Associate Director of Education, American Institute of Steel Construction.

DETERMINING BEAM FLEXURAL DESIGN STRENGTH FOR MINOR- AND MAJOR-AXIS BENDING

For a simply supported W24 $\times$ 76 beam, laterally braced only at the supports, determine the flexural design strength for (*a*) minor-axis bending and (*b*) major-axis bending. Use the Table 3-2 in Part 3 of the AISC *LRFD Manual*.

Calculation Procedure:

1. Determine if the beam is a compact section
The W24 $\times$ 76 is a compact section. This can be verified by noting that in the Properties Tables in Part 1 of the AISC *LRFD Manual*, both $b_f/2t_f$ and h_c/t_w for a W24 $\times$ 76 beam are less than the respective flange and web values of λ_p for $F_y = 50$ ksi (345 MPa).

2. Find the flexural design strength for minor-axis bending
For minor- (or *y*-) axis bending, $M_{ny} = M_{py} = Z_y F_y$, regardless of unbraced length. The available flexural design strength for minor-axis bending of a W24 $\times$ 76 is always equal to $\phi_b M_{ny} = \phi_b Z_y F_y = 0.90 \times 28.6$ in.$^3 \times 50$ ksi = 1287 kip-in. = 107.25 kip $\cdot$ ft (145 kNm).

3. Compute the available flexural design strength for major-axis bending
The flexural design strength for major-axis bending depends on C_b and L_b. For a simply supported member, the end moments $M_1 = M_2 = 0$; $C_b = 1.0$.

4. Plot the results

For $0 < L_b < (L_p = 8.0$ ft$)$, $\phi_b M_n = \phi_b M_p = 723$ kip · ft (980 kNm). At $L_b = L_r = 23.4$ ft, $\phi_b M_{nx} = \phi_b M_r = 342$ kip · ft (464 kNm). Linear interpolation is required for $L_p < L_b < L_r$. For $L_b > L_r$, refer to the beam graphs in Part 3 of the AISC *LRFD Manual*.

Figure 19b shows the data plotted for this beam, after using data from the AISC table referred to above.

Related Calculations. This procedure is the work of Abraham J. Rokach, MSCE, Associate Director of Education, American Institute of Steel Construction. SI values were prepared by the handbook editor.

DESIGNING WEB STIFFENERS FOR WELDED BEAMS

The welded beam in Fig. 20a (selected from the table of Built-Up Wide-Flange Sections in Part 3 of the AISC *LRFD Manual*) frames into the column in Fig. 20b. Design web stiffeners to double the shear strength of the web at the end panel.

Calculation Procedure:

1. Determine the nominal shear strength for a stiffened web

At the end panels there is no tension field action. The nominal shear strength for a stiffened web or unstiffened web according to the limits states of shear yielding and shear buckling is, using the AISC *LRFD Manual* equation, $V_n = 0.6 F_y A_w C_v$. Since this section is a built-up section,
When $h/t_w < 1.10 \sqrt{ek_v E/F_y}$, $C_v = 1.0$ and when $1.10 \sqrt{ek_v E/F_y} \le h/t_w \le 1.37 \sqrt{ek_v E/F_y}$
Then

$$C_v = \frac{1.10\sqrt{k_v E/F_y}}{h/t_w}$$

Assuming $h/t_w > 1.37 \sqrt{ek_v E/F_y}$ for this example then $C_v = \dfrac{1.51 k_v E}{(h/t_w)^2 F_y} V_n = 0.60 F_y A_w \dfrac{1.51 k_v E}{(h/t_w)^2 F_y}$.

In the case of no stiffeners $k_v = 5$.

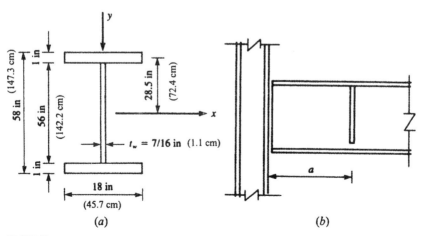

FIGURE 20

2. Check the original assumptions for doubling the shear strength
To double the shear strength, $k = 2 \times 5 = 10$. Then in AISC Eq. (G2-6)

$$k_v = 5 + \frac{5}{(a/h)^2} = 10$$

This implies $a/h = 1.0$ or $a = k$; thus, the clear distance between transverse web stiffeners $a = h = 56$ in. (142.2 cm). Checking the original assumption we obtain

$$\frac{h}{t_w} = \frac{56 \text{ in.}}{0.44 \text{ in.}} = 127.27 \geq 1.37\sqrt{k_v E/F_y} = 1.37 \times \sqrt{10 \times 29000 \text{ ksi}/50 \text{ ksi}} = 104.33.$$

Therefore OK.

3. Design the stiffener, trying a pair of stiffener plates
Stiffener design can be performed thusly. Because tension field action is not utilized, the equation $I_{st} \geq at_w^3 j$ must be satisfied, where

$$j = \frac{2.5}{(a/h)^2} - 2 \geq 0.5, \frac{2.5}{1^2} - 2 = 0.5, I_{st} \geq 56 \text{ in.} \times (0.44 \text{ in.})^3 \times 0.5 = 2.385 \text{ in.}^4 \ (99.28 \text{ cm}^4).$$

Try a pair of stiffener plates, 2.5×0.25 in. (6.35×0.635 cm). The moment of inertia of the stiffener pair about the web centerline

$$I_{st} = \frac{0.22 \text{ in.} \times (5.44 \text{ in.})^3}{12} = 3.35 \text{ in.}^4 > 2.34 \quad \text{OK} \quad (139.4 \text{ cm}^4 > 97.4 \text{ cm}^4) \text{ OK}$$

DETERMINING THE DESIGN MOMENT AND SHEAR STRENGTH OF A BUILT-UP WIDE-FLANGE WELDED BEAM SECTION

For the welded section in Fig. 20a (selected from the table of Built-Up Wide-Flange Sections in Part 3 of the AISC *LRFD Manual*), determine the design moment and shear strengths. Bending is about the major axis; $C_b = 1.0$. The (upper) compression flange is continuously braced by the 57 floor deck. Steel is A36.

Calculation Procedure:

1. Check the beam compactness and flange local buckling
Working with the "Compression Elements Members Subject to Flexure" Table B4.1b in the Appendix of the AISC *LRFD Specification*, the compactness of the beam (for a doubly symmetric I shape bending about its major axis) should first be checked:

Flange $\qquad \lambda = \dfrac{b}{t} = \dfrac{b_f}{2t_f} = \dfrac{18 \text{ in.}}{2 \times 1 \text{ in.}} = 9.0$

For the definition of b for a welded I shape, see the AISC *LRFD Manual* Table B4.1b

Flange $\qquad \lambda_p = 0.38eE/F_y = 9.15$

For the flange $\lambda < \lambda_p$. Therefore, the flange is compact and $M_{nx} = M_{px}$ for the limit state of flange local buckling (FLB).

STRUCTURAL STEEL DESIGN

Web $\quad\lambda = \dfrac{h}{t_w} = \dfrac{56 \text{ in.}}{7/16 \text{ in.}} = 128.0$

Web $\quad\lambda_p = 3.76\sqrt{\dfrac{E}{F_y}} = 3.76 \times \sqrt{\dfrac{29000}{36}} = 106.72$, therefore $\lambda < \lambda_p$

$\lambda_p = 5.70\sqrt{\dfrac{E}{F_y}} = 5.70 \times \sqrt{\dfrac{29000}{36}} = 161.78\ \lambda_p < \lambda < \lambda_r$ the web is therefore noncompact.

Hence $M_{rx} < M_{nx} < M_{px}$.

2. Analyze the lateral bracing relating to the limit state of lateral-torsional buckling (LTB)

For this continuously braced member $L_b = 0$; $M_{nx} = M_{px}$ for LTB. Summarizing:

Limit State	M_{nx}
LTB	M_{px}
FLB	M_{px}
WLB	$M_{rx} < M_{nx} < M_{px}$

The limit state of WLB (with minimum M_{nx}) governs. To determine M_{px}, M_{rx}, and M_{nx} for a doubly symmetric I-shaped member bending about the major axis, refer again to the AISC *LRFD Manual Table Sections F4 and F5*. Section F4 refers to doubly symmetric I-sections with compact or noncompact webs while Section F5 refers to doubly symmetric I-sections with slender webs. Use of F5 will produce more conservative results. For this example we will use section F4 where for:

Calculation Procedure:

1. Compression flange yielding, $M_{nx} = R_{pc}M_{yc} = R_{pc}F_y S_{xc}$, M_{yc} = yield moment, in kip-in. [Eq. (F4-1)]

2. Lateral torsional buckling,

$$M_{nx} = C_b\left[R_{pc}M_{yc} - (R_{pc}M_{yc} - F_L S_{xc})\left(\dfrac{L_b - L_p}{L_r - L_p}\right)\right] \leq R_{pc}M_{yc} \text{ when } L_p < L_b \leq L_r \quad \text{[Eq. (F4-2)]}$$

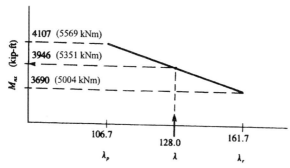

FIGURE 21

For $\lambda > \lambda_p$ but $< \lambda_r$

$$R_{pc} = \left[\frac{M_p}{M_{yc}} - \left(\frac{M_p}{M_{yc}} - 1\right)\left(\frac{\lambda - \lambda_p}{\lambda_r - \lambda_p}\right)\right] \le \frac{M_p}{M_{yc}} \qquad \text{[Eq. (F4-9b)]}$$

where R_{pc} is web plastification factor.
The properties S_x and Z_x of the cross section in Fig. 21 must now be calculated.

$$S_{xc} = I_x/c, \text{ where } c = d/2 = 58 \text{ in.}/2 = 29 \text{ in. (73.7 cm)}$$

The contributions of the two flanges and the web to the moment of inertia I_x are

Elements	$bt^3/12 + AD^2$
2 Flanges	[18 in. × (1 in.)^3/12 + (18 in. × 1 in.)(28.5 in.)^2]2 = 29,244 in.^{4} (1,217,227 cm^4)
Web	0.44 in. × (56 in.)^3/12 + 0 = 6403 in.^{4} (266,513 cm^4)
I_x	S_x = (29,244 + 6403)/29 = 1230 in.^{3} (20,156 cm^3)

To determine Z_x, we calculate: Σ_{AD}, where A is the cross-sectional area of each element and D represents its distance from the centroidal x-axis. In calculating Z_x, the upper and lower halves of the web are taken separately.

Elements	AD
2 Flanges	[(18 in. × 1 in.)(28.5 in.)]2 = 1026 in.^{3} (16,813 cm^3)
½ Web	[(28 in. × 0.44 in.) × 14 in.]2 = 343 in.^{4} (5620 cm^3)
	= 1369 in.^{3} (22,433 cm^3)
Z_x	Z_x = 1369 in.^{3} (22,433 cm^3)

3. Determine the welded section flexural strength
According to the AISC *Manual*, if $I_{yc}/I_y \le 0.23$ then $R_{PC} = 1.0$ AISC Eq. (F4-10). Let's assume this is the case for this example.
Determining flexural strengths, we obtain $M_p = F_y Z_x \le 1.6 F_y S_{xc}$.

$$M_{px} = 36 \text{ kips/in.}^2 \times 1369 \text{ in.}^3/(12 \text{ in./ft}) = 4107 \text{ kip} \cdot \text{ft (5568 kNm)}$$

$$M_{rx} = 36 \text{ kips/in.}^2 \times 1230 \text{ in.}^3/(12 \text{ in./ft}) = 3690 \text{ kip} \cdot \text{ft (5003 kNm)}$$

$1.6 M_{rx} = 1.6 \times 3690 \text{ kip} \cdot \text{ft} = 5904 \text{ kip} \cdot \text{ft (8004 kNm)}$ hence OK

The nominal shear strength, V_n, of unstiffened or stiffened webs based on the limit states of shear yielding and shear buckling is given by

$$V_n = 0.6 F_y A_w C_v, \text{ where } C_v, \text{ web shear coefficient, is given by}$$

i. When $h/t_w \le 1.10 \sqrt{ek_v E/F_y}$, then $C_v = 1.0$

ii. When $1.10\, ek_vE/F_y \leq h/t_w \leq 1.37\, ek_vE/F_y$, then $C_v = \dfrac{1.10\sqrt{k_vE/F_y}}{h/t_w}$

iii. When $h/t_w > 1.37\, ek_vE/F_y$, then $C_v = \dfrac{1.5k_vE}{(h/t_w)^2 F_y}$

Here $h/t_w = 56/0.44 = 127.27$. $k_v = 5 + \dfrac{5}{(a/h)^2}$; for $a/h > 3$ then $k_v = 5$. Assume $k_v = 5$ for this example, then $1.37e\,(5 \times 29000 \text{ ksi}/36 \text{ ksi}) = 87$. Therefore $h/t_w) = 128 > 1.37ek_vE/F_y = 87$. $C_v = 1.5 \times 5 \times 29{,}000/[(128)^{\wedge}2 \times 36] = 0.368$; therefore the nominal shear strength $V_n = 0.6 \times 36 \text{ kips/in.}^2 \times (56 \text{ in.} \times 44 \text{ in.}) \times 0.368 = 196 \text{ kips (872 kN)}$. And $\phi_v V_n = 0.9 \times 196 = 176.4 \text{ kips (785 kN)}$.

FINDING THE LIGHTEST SECTION TO SUPPORT A SPECIFIED LOAD

Find the lightest W8 in A992 Gr 50 steel to support a factored load of 100 kips (444.8 kN) in tension with an eccentricity of 6 in. (15.2 cm). The member is 6 ft (1.8 m) long and is laterally braced only at the supports; $C_b = 1.0$. Try the orientations (a) to (c) shown in Fig. 22.

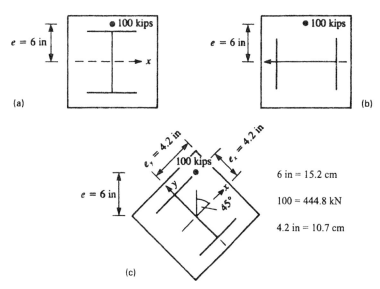

FIGURE 22

Calculation Procedure:

1. Try the first orientation, (a), Fig. 22
Given $P_u = 100$ kips (444.8 kN); $M_u = P_u e = 100$ kips $\times$ 6 in./(12 in./ft) = 50 kip $\cdot$ ft
For orientation (a) in Fig. 22

$$P_u = 100 \text{ kips} \quad M_{ux} = 50 \text{ kip} \cdot \text{ft} \quad M_{uy} = 0 \text{ kip} \cdot \text{ft}$$

Try a W8 $\times$ 28: the design tensile strength (for a cross section with no holes)

$$\phi_t P_n = \phi_t F_y A_g = 0.9 \times 50 \text{ ksi} \times 8.25 \text{ in.}^2 = 371 \text{ kips (1651 kN)}$$

For $(L_b = 6.0 \text{ ft}) < (L_p = 6.8 \text{ ft})$, the design flexural strength for x-axis bending $\phi_b M_{nx} = \phi_b Z_x F_y = 0.90 \times 27.2 \text{ in.}^3 \times 50 \text{ ksi}/(12 \text{ in./ft}) = 102$ kip $\cdot$ ft (138 kNm) which is also the tabulated value for $\phi_b M_p$ for a W8 $\times$ 28 in the Beam Selection Table 3-2 in Part 3 of the AISC LRFD *Manual*.
Since

$$\frac{P_u}{\phi_t P_n} = \frac{P_r}{P_c} = \frac{100 \text{ kips}}{371 \text{ kips}} = 0.27 > 0.2$$

Then $\dfrac{P_r}{P_c} + \dfrac{8}{9}\left(\dfrac{M_{rx}}{M_{cx}} + \dfrac{M_{ry}}{M_{cy}}\right) \leq 1.0$ AISC LRFD Manual Eq. (H1-1a)

$$0.27 + 8/9(50) \text{kip} \cdot \text{ft}/102 \text{ kip} \cdot \text{ft} + 0 = 0.70 < 1.0 \quad \text{OK.}$$

2. For the second orientation, (b), Fig. 22
Given $P_u = 100$ kips (444.8 kN); $M_u = P_u e = 100$ kips $\times$ 6 in./(12 in./ft) = 50 kip $\cdot$ ft
For orientation (b) in Fig. 22

$$P_u = 100 \text{ kips} \quad M_{ux} = 0 \text{ kip} \cdot \text{ft} \quad M_{uy} = 50 \text{ kip} \cdot \text{ft}$$

Try a W8 $\times$ 28: the design tensile strength (for a cross section with no holes)

$$\phi_t P_n = \phi_t F_y A_g = 0.9 \times 50 \text{ ksi} \times 8.25 \text{ in.}^2 = 371 \text{ kips (1651 kN)}$$

For all $L_b = 6.0$ ft, the design flexural strength for y-axis bending

$$\phi_b M_{ny} = \phi_b Z_y F_y = 0.90 \times 10.1 \text{ in.}^3 \times 50 \text{ ksi}/(12 \text{ in./ft}) = 38 \text{ kip} \cdot \text{ft (51.5 kNm)}.$$
Since $M_{uy} = 50$ kip $\cdot$ ft $>$

$\phi_b M_{ny} = 38$ kip $\cdot$ ft, a W8 $\times$ 28 is inadequate. Try W8 $\times$ 48: $A_g = 14.1$ in.^{2} (90 cm^2), $Z_y = 22.9$ in.^{3} (375.3 cm^3).

$$\phi_b M_{ny} = \phi_b Z_y F_y = (0.9 \times 22.9 \text{ in.}^3 \times 50 \text{ ksi})/12 \text{ in./ft} = 85.9 \text{ kip} \cdot \text{ft (116 kNm)}$$

$$\phi_t P_n = \phi_t F_y A_g = 0.9 \times 50 \text{ ksi} \times 14.1 \text{ in.}^2 = 634.5 \text{ kips (2822 kN)}$$

Because $(P_u/\phi_t P_n = P_r/P_c) = 100$ kips/634.5 kips = 0.16 < 0.2 AISC Eq. (H1-1b) applies. That is

$$\frac{P_r}{2P_c} + \left(\frac{M_{rx}}{M_{cx}} + \frac{M_{ry}}{M_{cy}}\right) \le 1.0$$

$0.16/2 + (0 + 50$ kip $\cdot$ ft/85.9 kip $\cdot$ ft$) = 0.66 < 1.0$ OK

3. Find the section for a load eccentric with respect to both principal axes
For orientation (c) in Fig. 22 assume that the load is eccentric with respect to both principal axes.
Referring to Fig. 22c

$e_x = e\cos 45° = 6$ in. $\times 0.7071 = 4.2$ in. (10.7 cm)

$e_y = e\sin 45° = 6$ in. $\times 0.7071 = 4.2$ in. (10.7 cm)

$M_{ux} = P_u e_x = 100$ kips $\times 4.2$ in./12 in./ft $= 35.4$ kip $\cdot$ ft (48 kNm)

$M_{uy} = P_u e_y = 100$ kips $\times 4.2$ in./12 in./ft $= 35.4$ kip $\cdot$ ft (48 kNm)

Again try a W8 × 48 as above

$(P_u/\phi_t P_n = P_r/P_c) = 100$ kips/634.5 kips $= 0.16 < 0.2$

$\phi_b M_{ny} = 85.9$ kip $\cdot$ ft (116 kNm)

From AISC Manual Table 3-2, $\phi_b M_{nx} = 184$ kip $\cdot$ ft (249.5 kNm), therefore in Interaction formula H1-1b:

$0.16/2 + [(35.4/184) + (35.4/85.9)] = 0.68 < 1.0$ OK

The most efficient configuration is orientation (a), strong axis bending, which requires a W8 × 28 as opposed to a W8 × 48 for the other two cases.

COMBINED FLEXURE AND COMPRESSION IN BEAM-COLUMNS IN A BRACED FRAME

Select, in A992 Gr. 50 steel, a W14 section for a beam-column in a braced frame with the following combination of factored loads: $P_u = 800$ kips (3558 kN); first-order moments $M_{ux} = 200$ kip $\cdot$ ft (271 kNm); $M_{uy} = 0$; single-curvature bending (i.e., equal and opposite end moments); and no transverse loads along the member. The floor-to-floor height is 15 ft (4.57 m).

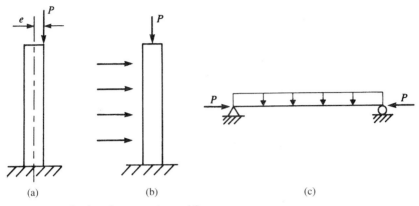

FIGURE 23. Combined compression and flexure.

Calculation Procedure:

1. Find the effective axial load for the beam-column

This procedure considers singly and doubly symmetric beam-columns: members subjected combined axial compression and bending about one or both principal axes. The combination of compression with flexure may result from (either).

   a. A compressive force that is eccentric with respect to the centroidal axis of the column, as in Fig. 23a
   b. A column subjected to lateral force or moment, as in Fig. 23b
   c. A beam transmitting wind or other axial forces, as in Fig. 23c

Interaction Formulas:

The cross sections of beam-columns must comply with formula (H1-1a) or (H1-1b), whichever is applicable.

For $(P_r/P_c) \geq 0.2$, then

$$\frac{P_r}{P_c} + \frac{8}{9}\left(\frac{M_{rx}}{M_{cx}} + \frac{M_{ry}}{M_{cy}}\right) \leq 1.0$$

For $(P_r/P_c) < 0.2$, then

$$\frac{P_r}{2P_c} + \left(\frac{M_{rx}}{M_{cx}} + \frac{M_{ry}}{M_{cy}}\right) \leq 1.0$$

where

P_r = required axial strength, kips (kN)
$P_c = \phi_c P_n$ = available axial strength, kips (kN)
M_r = required flexural strength, kip · ft (kNm)
$M_c = \phi_b M_n$ = available flexural strength, kip · ft (kNm)
ϕ_c = resistance factor for compression = 0.90
ϕ_b = resistance factor for flexure = 0.90

The subscript x refers to bending about the major principal centroidal (or x) axis; y refers to the minor principal centroidal (or y) axis.

SIMPLIFIED SECOND-ORDER ANALYSIS

Second-order moments in beam-columns are the additional moments caused by the axial compressive forces acting on a displaced structure. Normally, structural analysis is first-order; that is, the everyday methods used in practice (whether done manually or by one of the popular computer programs) assume the forces as acting on the original undeflected structure. Second-order effects are neglected. To satisfy the AISC LRFD Specification, second-order moments in beam-columns must be considered in their design.

Instead of rigorous second-order analysis, the AISC LRFD Specification presents a simplified alternative method. The components of the total factored moment determined from a first-order elastic analysis (neglecting secondary effects) are divided into two groups, M_{nt} and M_{lt}.

1. M_{nt}—the required flexural strength in a member assuming there is no lateral translation of the structure. It includes the first-order moments resulting from the gravity loads (i.e., dead and live loads), calculated manually or by computer.

2. M_{lt}—the required flexural strength in a member due to lateral frame translation. In a braced frame, $M_{lt} = 0$. In an unbraced frame, M_{lt} includes the moments from the lateral loads. If both the frame and its vertical loads are symmetric, M_{lt} from the vertical loads is zero. However, if either the vertical loads (i.e., dead and live loads) or the frame geometry is asymmetric and the frame is not braced, lateral translation occurs and $M_{lt} \neq 0$. To determine M_{lt} (*a*) apply fictitious horizontal reactions at each floor level to prevent lateral translation and (*b*) use the reverse of these reactions as "sway forces" to obtain M_{lt}. This procedure is illustrated in Fig. 24. As is indicated there, M_{lt} for an unbraced frame is the

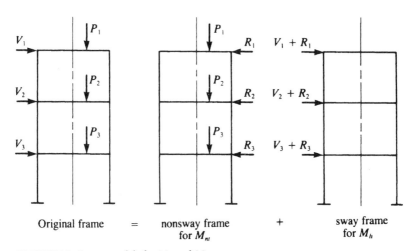

FIGURE 24. Frame models for M_{lt} and M_{nt}.

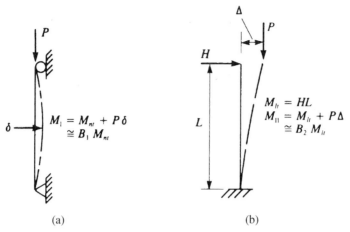

FIGURE 25. Illustrations of secondary effects. (*a*) Column in braced frame; (*b*) Column in unbraced frame.

sum of the moments due to the lateral loads and the "sway forces." Once M_{nt} and M_{lt} have been obtained, they are multiplied by their respective magnification factors, B_1 and B_2, and added to approximate the actual second-order factored moment M_r.

$$M_r = B_1 M_{nt} + B_2 M_{lt}$$

$$P_r = P_{nt} + B_2 P_{lt}$$

As shown in Fig. 25, B_1 accounts for the secondary $P - d$ effect in all frames (including sway inhibited), and B_2 covers the $P - \Delta$ effect in unbraced frames. The analytical expressions for B_1 and B_2 follow:

$$B_1 = \frac{C_m}{1 - \alpha P_r/P_{e1}} \geq 1.0$$

where P_r is the factored axial compressive force in the member, kips and P_{e1} is elastic critical buckling strength of the member in the plane of bending, calculated based on the assumption of no lateral translation at the member ends, kips (kN) and is given by

$$P_{e1} = \frac{\pi^2 EI}{K_1 L}$$

EI^* = flexural rigidity required to be used in the analysis ($= 0.8 \tau_b EI$ when used in the direct analysis method where τ_b is as defined in Chapter C of the AISC Manual; $= EI$ for the effective length and first-order analysis methods)

E = modulus of elasticity of steel = 29,000 ksi (200,000 MPa)

where $K = 1.0$, I is the moment of inertia (in.^{4}) (cm^4) and l is the unbraced length (in.) (cm) (Both I and l are taken in the plane of bending only.)

The coefficient C_m is determined as follows.

(1) For restrained beam-columns not subjected to transverse loads between their supports in the plane of bending

$$C_m = 0.6 - 0.4(M_1/M_2)$$

where M_1/M_2 is the ratio of the smaller to larger moment at the ends of the portion of the member unbraced in the plane of bending under consideration. If the rotations due to end moments M_1 and M_2 are in opposite directions, then M_1/M_2 is negative; otherwise M_1/M_2 is positive.

(2) For beam-columns subjected to transverse loads between supports, if the ends are restrained against rotation, $C_m = 0.85$; if the ends are *unrestrained* against rotation, $C_m = 1.0$.

Two equations are given for B_2 in the AISC *LRFD Specification*:

$$B_2 = \frac{1}{1 - \dfrac{\alpha P_{story}}{P_{e,story}}} \geq 1.0$$

where $\alpha = 1$
  P_{story} = total vertical load supported by the story, including loads in columns that are not part of the lateral force resisting system, kips (kN)
  $P_{e,story}$ = elastic critical buckling strength for the story in the direction of translation being considered, kips (kN), determined by side-sway buckling analysis or as:
  $P_{e,story} = RM(HL/\Delta_H)$
  $RM = 1 - 0.15(P_{mf}/P_{story})$
  L = height of story, in.
  P_{mf} = total vertical load in columns in the story that are part of moment frames, if any, in the direction of translation being considered (= 0 for braced frame systems), kips (kN)
  Δ_H = first-order interstory drift, in the direction of translation being considered, due to lateral forces, in. (cm), computed using the stiffness required to be used in the analysis (stiffness reduced as provided in Section C2.3 of the AISC *Manual* when the direct analysis method is used). Where Δ_H varies over the plan area of the structure, it shall be the average drift weighted in proportion to vertical load or, alternatively, the maximum drift.
  H = story shear, in the direction of translation being considered, produced by the lateral forces used to compute Δ_H, kips (kN)

Often, especially for tall buildings, the maximum drift index is a design criterion. For columns with biaxial bending in frames unbraced in both directions, two values of B_1 (B_{1x} and B_{1y}) are needed for each column and two values of B_2 for each story, one for each major direction.

Once the appropriate B_1 and B_2 have been evaluated, Eq. (A-8-1) can be used to determine M_{rx} and M_{ry} for the applicable interaction formula.

$P_u = 800$ kips (3558 kN); $M_{ux} = 200$ kip · ft (271 kNm); $M_{uy} = 0$; $KL_x = 15$ ft (4.57 m); $KL_y = 15$ ft (4.57 m). ASTM A992, $F_y = 50$ ksi; $F_u = 65$ ksi; values p, b_x, b_y, t_y, t_r are calculated as follows:

Axial Compression, $p = 1/\phi_c P_n$ (kips^{-1})

Strong axis bending, $b_x = 8/(9\phi_b M_{nx})$ (kip · ft)$^{-1}$

Weak axis bending, $b_y = 8/(9\phi_b M_{ny})$ (kip · ft)$^{-1}$

Tension yielding, $t_y = 1/(\phi_t F_y A_g)$ (kips)$^{-1}$

Tension rupture, $t_r = 1/(\phi_t F_u(0.75A_g))$ (kips)$^{-1}$. Try W14 × 159. From Table 6-1 of the AISC *Manual* for W14 × 159,

$p = 0.553/10^3$ (kips^{-1}); $b_x = 0.831/10^3$ (kip · ft)$^{-1}$; $b_y = 1.62/10^3$ (kip · ft)$^{-1}$

From Eqs. (6-1) and (6-2) of the AISC *Manual*;
When $pP_r \geq 0.2$
Then

$$pP_r + b_x M_{rx} + b_y M_{ry} \leq 1.0 \quad \text{(AISC Eq. 6-1)}$$

When $pP_r < 0.2$

$$\frac{1}{2}pP_r + \frac{9}{8}(b_x M_{rx} + b_y M_{ry}) \leq 1.0 \quad \text{(AISC Eq. 6-2)}$$

$$pP_r = (0.553/1000) \times 800 = 0.442 > 0.2$$

Therefore

$pP_r + b_x M_{rx} + b_y M_{ry} = 0.442 + (0.831/1000) \times 200 + (1.62/1000) \times 0 = 0.61 < 1.0 \quad \text{OK}$

2. Analyze the braced frame
To determine second order moments use Eq. (A-8-1)

$$M_r = B_1 M_{nt} + B_2 M_{lt}$$

Because the frame is braced $M_{lt} = 0$, therefore $M_r = B_1 M_{nt} = B_1 \times 200$ kip · ft, but B_1 is given by

$$B_1 = \frac{C_m}{1 - \alpha^{P_r}/P_{e1}} \geq 1.0$$

$$C_m = 0.6 - 0.4(M_1/M_2)$$

$$M_1/M_2 = -200/200 = -1.0$$

$$C_m = 0.6 - 0.4(-1.0) = 1.0$$

Moment of Inertia of a W14 × 159, $I_x = 1900$ in.^{4} (79,084 cm^4)

$P_{e1} = \dfrac{\pi^2 EI}{K_1 L}$; $P_{e1} = [(3.14)^2 \times 29{,}000 \text{ kips/in.}^2 \times 1900 \text{ in.}^4]/(1.0 \times 15 \text{ ft} \times 12 \text{ in./ft}) = 16{,}784$ kips
(74661 kN), therefore $B_1 = 1.0/(1 - (1.0 \times (800/16784))) = 1.05$
Here, $M_{ux} = 1.05 \times 200$ kip · ft = 210 kip · ft (284.6 kNm) the second-order required flexural strength.

3. Determine flexural strength
The flexural strengths for W14 × 159 for both limit states are provided in the AISC LRFD *Manual* as:
$\phi_b M_{px} = 1080$ kip · ft for $L_p = 14.1$ ft and $\phi_b M_{nx} = 667$ kip · ft for $L_r = 66.7$ ft, respectively.
Since $L = 15$ ft > 14.1 therefore available flexural strength, $M_{cx} = 667$ kip · ft, substituting into the interaction equation $\dfrac{P_r}{P_c} + \dfrac{8}{9}\left(\dfrac{M_{rx}}{M_{cx}} + \dfrac{M_{ry}}{M_{cy}}\right) \leq 1.0$ gives

$$0.442 + 8/9(210/667) + 0 = 0.72 < 1.0, \text{ therefore OK.}$$

By a similar solution of interaction formula (*H1-1a*), it can be shown that a W14 × 145 is also adequate.

SELECTION OF CONCRETE-FILLED STEEL COLUMN

Select a 6-in. (15.2-cm) concrete-filled steel-pipe column for a required axial compressive strength of 200 kips (889.6 kN), where $KL = 10.0$ ft (3.05 m), $F_y = 35$ ksi (241 MPa), $f'_c = 3.5$ ksi (24.1 MPa), using normal-weight concrete = 145 lb/ft³ (2320 kg/m³).

Calculation Procedure:

1. Try a standard-weight concrete-filled pipe
2. Analyze the selected column
Check minimum wall thickness of pipe, Fig. 26. From the AISC *Manual* Table I1.1a for a Round HSS compression member subject to axial compression the following width-thickness-ratio should be satisfied assuming section is compact:

$$D/t \le 0.15E/F_y$$

The minimum thickness shall be $t \ge D\,(1/0.15E/F_y) = D(6.67F_y/E) = 6 \times (6.67 \times 35\text{ ksi})/29000\text{ ksi} = 0.048$ in. Use $t = 0.25$ in.

Check minimum cross-sectional area of steel pipe

$$A_s = \pi(R^2 - R_i^2) = \pi(D^2 - D_i^2)/4 = 3.14/4(6.5^2 - 6^2) = 4.908 \text{ in.}^2 \ (31.67 \text{ cm}^2)$$

Area of concrete $A_c = \pi D_i^2/4 = 3.14(6)^2/4 = 28.26 \text{ in.}^2 (182.32 \text{ cm}^2)$

Check percentage of steel: $\dfrac{A_s}{A_s + A_c} = 4.908/(4.908 + 28.26) = 0.15 > 1\%$ OK.

3. Analyze the selected column
Since the section is compact, i.e., $D/t = 6.5/.25 = 26 < 0.15E/F_y = 124$. Therefore $P_{no} = P_p$, where $P_p = F_y A_s + C_2 f'_c (A_c - A_{sr} E_s/E_c)$ AISC *Manual* Eq. (I2-9b). In this example there are no reinforcing bars, therefore $A_{sr} = 0$ and $C_2 = 0.95$ for round sections.

$P_p = 35$ ksi × 4.908 in.² + 0.95 × 3.5 ksi × 28.26 in.² − 0 = 265.74 kips (1182 kN) > 200 kips (889 kN).

For noncompact and slender members follow procedures in section I2.2b of the AISC *Manual*.

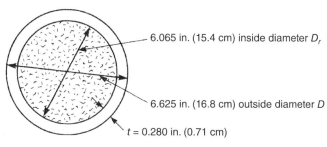

FIGURE 26

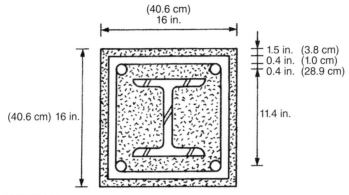

FIGURE 27

DETERMINING DESIGN COMPRESSIVE STRENGTH OF COMPOSITE COLUMNS

Determine the design compressive strength of a W8 × 40, Grade 50 steel encased in a 16 × 16-in. (40.6 × 40.6 cm) (f'_c = 3.5 ksi) (24.1 MPa) normal-weight concrete column in Fig. 27.

Reinforcement is four No. 7 (Grade 60) bars longitudinally, and No. 3 ties at 10 in. (25.4 cm) horizontally.

Calculation Procedure:

1. Check the minimum requirements for the column
Checking minimum requirements

(a) For a W8 × 40, A_s = 11.7 in.^{2}, total area = 16 × 16 in. = 256 in.^{2} (1652 cm^2)
Area ratio = 11.7 in.^{2}/256 in.^{2} = 4.6% > 1%

(b) Lateral tie spacing = 8 in. (20.32 cm) ≤ 0.5 × 16 in. least dimension = 8 in. (20.32 cm)

(c) Lateral ties No. 3 bars: A_r = 0.11 in.^{2} (71.26 mm^2)

(d) Minimum clear cover = 1.5 in. (3.8 cm) OK

(e) Minimum longitudinal reinforcement ratio, $A_{sr}/A_g \geq 0.004$
$A_{sr} = 4 \times \pi d^2/4 = 4 \times (0.875)^2 \times 3.14/4 = 2.405$ in.^{2} (15.5 cm^2)
2.405 in.^{2}/256 in.^{2} = 0.061 > 0.004 OK

(f) Maximum longitudinal reinforcement ratio, A_{sr}/A_g = 0.061 ≤ 0.08, therefore OK.

2. Determine available compressive strength
The nominal available compressive strength P_{no} without consideration of length effects is given by

$P_{no} = F_y A_s + F_{yr} A_{sr} + 0.85 f'_c A_c$ = (50 ksi) (11.7 in.^{2}) + (60 ksi)(2.405 in.^{2}) + 0.85 (3.5 ksi) (241.9 in.^{2}) = 1,448.94 kips (6445 kN)

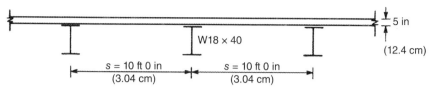

FIGURE 28

ANALYZING A CONCRETE SLAB FOR COMPOSITE ACTION

A W18 × 40 interior beam is shown in Fig. 28. Steel is Gr. 50, beam span is 30 ft 0 in. (9.14 m), and beam spacing 10 ft 0 in. (3.04 m). The beams are to act compositely with a 5-in. (12.7-cm) normal weight concrete slab; $f'_c = 5.0$ ksi (34.5 MPa). Determine: (a) The effective width of concrete slab for composite action; (b) V_h (the total horizontal shear force to be transferred) for full composite action; (c) the number of 0.75-in. (1.9-cm) diameter shear studs required if $F_u = 65$ ksi (448 MPa).

Calculation Procedure:

1. Find the effective width of concrete slab for composite action
For an interior beam, the effective slab width on either side of the beam centerline is the minimum of

$$L/8 = 30.0 \text{ ft}/8 = 3.75 \text{ ft} = 45 \text{ in. (114.3 cm)} \textbf{ controls}$$

$$s/2 = 10 \text{ ft}/2 = 5 \text{ ft} = 60 \text{ in. (1.52 m)}$$

The effective slab width is therefore 2 × 45 in. = 90 in. (228.6 cm)

2. Determine the total horizontal shear force for full composite action
In positive moment regions, V_h for full composite action is the smaller of

$$0.85 f'_c A_c = 0.85 \, (5 \text{ ksi}) \, (90 \times 5 \text{ in.}) = 1913 \text{ kips (8509 kN)}$$

$$A_s F_y = 11.8 \text{ in.}^2 \times 50 \text{ ksi} = 590 \text{ kips (2624 kN)}$$

$$V_h = 590 \text{ kips}$$

3. Find the number of shear studs required
The nominal strength of a single shear stud

$$Q_n = 0.5 A_{sc} \sqrt{f'_c E_c} \leq A_{sc} F_u$$

For a ¾-in. diameter stud,

$A_{sc} = \pi(0.75)^2/4 = 0.44$ in.^{2} (2.84 cm^2); $E_c = w^{1.5} \sqrt{f'_c} = 145^{1.5} \sqrt{5.0 \text{ ksi}} = 3904$ ksi (26,917 MPa)
$F_u = 65$ ksi (448 MPa)

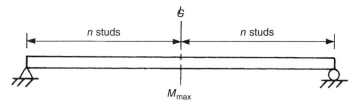

FIGURE 29

Therefore $Q_n = 0.5(0.44 \text{ in.}^2) \, e(5 \text{ ksi})(3904 \text{ ksi}) = 30.9 \text{ kips} \leq (0.44 \text{ in.}^2)(65 \text{ ksi}) = 30.9 \text{ kips} < 28.6 \text{ kips}$; therefore $Q_n = 28.6$ kips.

The number of shear connectors between the points of zero and maximum moments is

$$n = V_h/Q_n = 590 \text{ kips}/28.6 \text{ kips} = 21 \text{ studs}$$

For the beam shown in Fig. 29 the required number of shear studs is $2n = 2 \times 21 = 42$.

Assuming a single line of shear studs (over the beam web), stud spacing = 30.0 ft/42 = 0.71 ft = 8.5 in. (21.8 cm). This is greater than the six-stud diameter [or $6 \times \frac{3}{4}$ in. = 4.5 in. (11.4 cm)] minimum spacing, and less than the eight slab thickness [or 8×5 in. = 40 in. (101.6 cm)] maximum spacing, which is satisfactory.

DETERMINING THE DESIGN SHEAR STRENGTH OF A BEAM WEB

The end of a W12 × 86 beam (A992 steel) has been prepared as shown in Fig. 30 for connection to a supporting member. The three holes are 15/16 in. (2.38 cm) in diameter for ⅞-in. (2.22-cm) diameter bolts. Determine the design shear strength of the beam web.

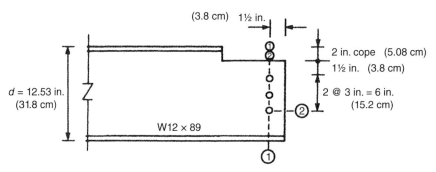

FIGURE 30

Calculation Procedure:

1. Find the applicable limit states
The applicable limit states are shear yielding, shear fracture, and block shear rupture.

For shear yielding [of gross section (1) in Fig. 30]. The nominal shear strength, V_n, based on shear yielding or shear buckling is given by $V_n = 0.6F_y A_w C_v$, $\phi_v = 0.90$ but most W shapes are not subject to shear buckling therefore

$\phi_v = 1.00$ for all W, S, HP shapes

$A_w = $ (d-cope) $\times t = (12.53 - 2$ in.$) \times 0.515$ in. $= 5.42$ in.^{2} (34.96 cm^2)

$\phi_v V_n = 1.0 \times 0.60 \times 5.42$ in.$^2 \times 50$ ksi $= 162$ kips (723 kN)

For shear fracture [of net section (1) of Fig. 11-9], $\phi_v V_n = 0.75 \times 0.60 \times A_{ns} \times F_u$, where $\phi_v = 0.75$, $F_u = 65$ ksi.

$A_{ns} = $ (d-cope-$3d_h$) $\times t = (12.53 - 2$ in. $- 3 \times 15/16) \times 0.515 = 3.97$ in.^{2} (25.6 cm^2)

$\phi_v V_n = 0.75 \times 0.60 \times 3.97$ in.$^2 \times 65$ ksi $= 116$ kips (516 kN)

For Block shear rupture [of section (2) in Fig. 11-9], $\phi_v = 0.75$, and $V_n = $ is the larger of the following:

$$0.60 A_{vg} F_y + A_n F_u \quad \text{or} \quad 0.6 A_{ns} F_u + A_g F_y$$

where $A_{vg} = $ gross area of the vertical part of (2)
$A_{ns} = $ net area of the vertical part of (2)
$A_g = $ gross area of the horizontal part of (2)
$A_n = $ net area of the horizontal part of (2)
$A_w = (1\frac{1}{2}$ in.$^2 \times 3$ in.$) \times 0.515$ in. $= 3.86$ in.^{2} (24.9 cm^2)
$A_{ns} = (1\frac{1}{2}$ in. $+ 2 \times 3$ in. $\times 2\frac{1}{2} \times 15/16) \times 0.515$ in. $= 2.66$ in.^{2} (17.2 cm^2)
$A_g = 1\frac{1}{2}$ in. $\times 0.515$ in. $= 0.77$ in.^{2} (4.96 cm^2)
$A_n = (1\frac{1}{2}$ in. $- \frac{1}{2} \times 15/16$ in.$) \times 0.515$ in. $= 0.53$ in.^{2} (3.42 cm^2)

2. Determine the design shear strength
V_n is the greater of

$$0.60 \times 3.86 \text{ in.}^2 \times 50 \text{ ksi} + 0.53 \text{ in.}^2 \times 65 \text{ ksi} = 150.25 \text{ kips (668 kN)}$$

and

$$0.60 \times 2.66 \text{ in.}^2 \times 65 \text{ ksi} + 0.77 \text{ in.}^2 \times 50 \text{ ksi} = 142.24 \text{ kips (633 kN)}$$

$$\phi V_n = 0.75 \times 150.25 = 112.69 \text{ kips (501 kN)}$$

The design shear strength is 113 kips (501 kN), based on the governing limit state of block shear rupture.

DESIGNING A BEARING PLATE FOR A BEAM AND IT'S END REACTION

The unstiffened end of a W21 × 62 beam in A992 steel rests on a concrete support ($f'_c = 3$ ksi) [20.7 MPa], Fig. 31. Design a bearing plate for the beam and its (factored) end reaction of 100 kips (444.8 kN). Assume the area of concrete support $A_2 = 6 \times A_1$ (the area of the bearing plate).

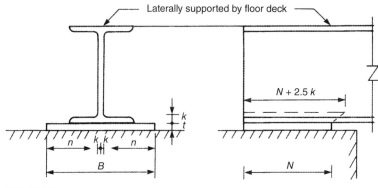

FIGURE 31

Calculation Procedure:

1. Find the bearing length

For the concentrated compressive reaction of 100 kips (444.8 kN) acting on the bottom flange, the applicable limit states are (1) local web yielding and (2) web crippling. (It is assumed that the beam is welded to the base plate and both are anchor-bolted to the concrete support. This should provide adequate lateral bracing to prevent side-sway web buckling.)

Corresponding to the applicable limit states are Eqs. (K1-3) and (K1-5), each of which has N, the length of bearing, as a parameter.

Solving for N, we obtain:

For Web Local Yielding; $R_u \leq \phi R_n = \phi(2.5k + N)F_t t_w$ (AISC Eq. J10-3)

100 kips ≤ 1.0 × (2.5 × 1⅜ in. + N) × 50 kips/in.² × 0.40 in. (1.01 cm)

$N \geq 1.562$ in. (3.967 cm)

For web crippling; $\phi = 0.75$; $R_u \leq \phi R_n = 0.8 t_w^2 \left[1 + 3\left(\frac{N}{d}\right)\left(\frac{t_w}{t_f}\right)^{1.5}\right]\sqrt{\frac{EF_{yw} t_f}{t_w}}$ (AISC Eq. J10-4)

100 kips = 0.8 × (0.4)²[1 + 3(N in./21 in.)(0.4/0.615)^{1.5}]

× e (29000 ksi × 50 ksi × 0.615 in.)/(0.40 in.) $N \geq 4.16$ in.

The minimum length of bearing is $N = 4.16$ in. (10.57 cm). Rounding up to the next full inch, let $N = 5$ in. (12.7 cm).

2. Compute the area of the bearing plate

The area of the bearing plate is determined by the bearing strength of the concrete support. Using the following equation, the design bearing strength is $\phi_c P_p$, where $\phi_c = 0.65$ and $P_p = 0.85 f'_c A_1 e (A_2/A_1)$, where $e(A_2/A_1) \leq 2.0$ then

100 kips = 0.65 × 0.85 × 3 ksi × A_1 × 2

Therefore the area of the bearing plate is $A_1 = 30.2$ in.² (194.6 cm²)

Because the bearing plate dimensions are $BN \geq A_1$:

$$B \geq A_1/N = 30.2 \text{ in.}^2/5 \text{ in.} = 6 \text{ in. (15 cm)}$$

However, B cannot be less than the flange width of the W21 × 62 beam, $b_f = 8.24$ in. Rounding up, let $B = 9$ in. (22.9 cm). A formula for bearing plate thickness is given by

$$t = \sqrt{\frac{2.22 R n^2}{A_1 F_y}}, \text{ where } R = 100 \text{ kips (444.8 kN)}$$

$$n = (B - 2k)/2 = (9 \text{ in.} - 2 \times 1\tfrac{3}{8} \text{ in.})/2 = 3.13 \text{ in. (7.95 cm)}$$

$$A_1 = BN = 9 \text{ in.} \times 5 \text{ in.} = 45 \text{ in.}^2 \text{ (290 cm}^2\text{)}$$

3. Select bearing plate dimensions

$$t = \sqrt{\frac{2.22 \times 100 \text{ kips} \times 3.13^2}{45 \text{ in.}^2 \times 50 \text{ ksi}}} = 0.983 \text{ in. (2.497 cm)}$$

Use a bearing plate size of $1 \times 5 \times 9$ in.

DETERMINING BEAM LENGTH TO ELIMINATE BEARING PLATE

Determine if the bearing plate chosen in the preceding procedure can be eliminated by altering the design.

Calculation Procedure:

1. Compute the needed thickness of the bottom flange

For the W21 × 62 beam to bear directly on the concrete support, its bottom flange must be sufficiently thick to act as a bearing plate. Let

$$t = \sqrt{\frac{2.22 R n^2}{A_1 F_y}} = 0.615 \text{ in. (1.56 cm)}$$

the flange thickness of the W21 × 62 beam. Because $B = b_f = 8.24$ in. (20.9 cm)

$$n = (B - 2k)/2 = (8.24 \text{ in.} - 2 \times 1\tfrac{3}{8} \text{ in.})/2 = 2.75 \text{ in. (6.73 cm)}$$

$$t = \sqrt{\frac{2.22 \times 100 \text{ kips} \times 2.75^2}{A_1 \text{ in.}^2 \times 50 \text{ ksi}}} = 0.615 \text{ in.}$$

2. Find the required length of bearing of the beam

From the above Eq. $A_1 = 89$ in.^{2} (533 cm^2) > 30.2 in.^{2} required for bearing on concrete.

$$N = A_1/B = A_1/b_f = 89 \text{ in.}^2/8.24 \text{ in.} = 10.8 \text{ in. (26.5 cm)}$$

By increasing the length of bearing of the beam to 11 in. (26.95 cm), the bearing plate can be eliminated.

Analysis of Stress and Strain

The notational system for axial stress and strain used in this section is as follows: A = cross-sectional area of a member; L = original length of the member; Δl = increase in length; P = axial force; s = axial stress; ε = axial strain = $\Delta l/L$; E = modulus of elasticity of material = s/ε. The units used for each of these factors are given in the calculation procedure. In all instances, it is assumed that the induced stress is below the proportional limit. The basic stress and elongation equations used are $s = P/A$; $\Delta l = sL/E = PL/(AE)$. For steel, $E = 30 \times 10^6$ lb/sq.in. (206 GPa).

STRESS CAUSED BY AN AXIAL LOAD

A concentric load of 20,000 lb (88,960 N) is applied to a hanger having a cross-sectional area of 1.6 sq.in. (1032.3 mm²). What is the axial stress in the hanger?

Calculation Procedure:

1. Compute the axial stress
Use the general stress relation $s = P/A = 20,000/1.6 = 12,500$ lb/sq.in. (86,187.5 kPa).

Related Calculations. Use this general stress relation for a member of any cross-sectional shape, provided the area of the member can be computed and the member is made of only one material.

DEFORMATION CAUSED BY AN AXIAL LOAD

A member having a length of 16 ft (4.9 m) and a cross-sectional area of 2.4 sq.in. (1548.4 mm²) is subjected to a tensile force of 30,000 lb (133.4 kN). If $E = 15 \times 10^6$ lb/sq.in. (103 GPa), how much does this member elongate?

Calculation Procedure:

1. Apply the general deformation equation
The general deformation equation is $\Delta l = PL/(AE) = 30,000(16)(12)/[2.4(15 \times 10^6)1 = 0.16$ in. (4.06 mm).

Related Calculations. Use this general deformation equation for any material whose modulus of elasticity is known. For composite materials, this equation must be altered before it can be used.

DEFORMATION OF A BUILT-UP MEMBER

A member is built up of three bars placed end to end, the bars having the lengths and cross-sectional areas shown in Fig. 32. The member is placed between two rigid surfaces and axial loads of 30 kips (133 kN) and 10 kips (44 kN) are applied at A and B, respectively. If $E = 2000$ kips/sq.in. (13,788 MPa), determine the horizontal displacement of A and B.

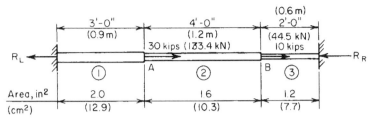

FIGURE 32

Calculation Procedure:

1. Express the axial force in terms of one reaction
Let R_L and R_R denote the reactions at the left and right ends, respectively. Assume that both reactions are directed to the left. Consider a tensile force as positive and a compressive force as negative. Consider a deformation positive if the body elongates and negative if the body contracts.

Express the axial force P in each bar in terms of R_L because both reactions are assumed to be directed toward the left. Use subscripts corresponding to the bar numbers (Fig. 32). Thus, $P_1 = R_L$, $P_2 = -30$; $P_3 = R_L - 40$.

2. Express the deformation of each bar in terms of the reaction and modulus of elasticity
Thus, $\Delta l_1 = R_L(36)/(2.0E) = 18RL/E$; $\Delta l_2 = (R_L - 30)(48)/(1.6E) = (30RL - 900)/E$; $\Delta l_3 = (R_L - 40)24/(1.2E) = (20R_L - 800)/E$.

3. Solve for the reaction
Since the ends of the member are stationary, equate the total deformation to zero, and solve for R_L. Thus $\Delta l_t = (68R_L - 1700)/E = 0$; $R_L = 25$ kips (111 kN). The positive result confirms the assumption that R_L is directed to the left.

4. Compute the displacement of the points
Substitute the computed value of R_L in the first two equations of step 2 and solve for the displacement of the points A and B. Thus $\Delta l_1 = 18(25)/2000 = 0.225$ in. (5.715 mm); $\Delta l_2 = [30(25) - 900]/2000 = -0.075$ in. (−1.905 mm).

Combining these results, we find the displacement of $A = 0.225$ in. (5.715 mm) to the right; the displacement of $B = 0.225 - 0.075 = 0.150$ in. (3.81 mm) to the right.

5. Verify the computed results
To verify this result, compute R_R and determine the deformation of bar 3. Thus $\Sigma F_H = -R_L + 30 + 10 - R_R = 0$; $R_R = 15$ kips (67 kN). Since bar 3 is in compression, $\Delta l_3 = -15(24)/[1.2(2000)] = -0.150$ in. (−3.81 mm). Therefore, B is displaced 0.150 in. (3.81 mm) to the right. This verifies the result obtained in step 4.

REACTIONS AT ELASTIC SUPPORTS

The rigid bar in Fig. 33a is subjected to a load of 20,000 lb (88,960 N) applied at D. It is supported by three steel rods, 1, 2, and 3 (Fig. 33a). These rods have the following relative cross-sectional areas: $A_1 = 1.25$, $A_2 = 1.20$, $A_3 = 1.00$. Determine the tension in each rod caused by this load, and locate the center of rotation of the bar.

1.64 STRUCTURAL ENGINEERING

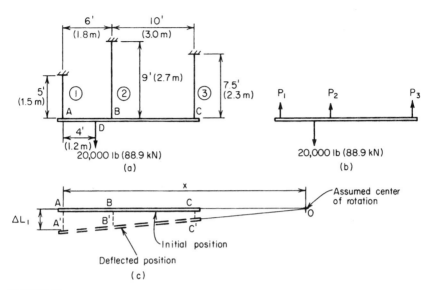

FIGURE 33

Calculation Procedure:

1. Draw a free-body diagram; apply the equations of equilibrium

Draw the free-body diagram (Fig. 33b) of the bar. Apply the equations of equilibrium: $\Sigma F_V\ P_1 + P_2 + P_3 - 20,000 = 0$, or $P_1 + P_2 + P_3 = 20,000$, Eq. a; also, $\Sigma M_C = 16P_1 + 10P_2 - 20,000(12) = 0$, or $16P_1 + 10P_2 = 240,000$, Eq. b.

2. Establish the relations between the deformations

Selecting an arbitrary center of rotation O, show the bar in its deflected position (Fig. 33c). Establish the relationships among the three deformations. Thus, by similar triangles, $(\Delta l_1 - \Delta l_2)/(\Delta l_2 - \Delta l_3) = 6/10$, or $10\Delta l_1 - 16\Delta l_2 + 6\Delta l_3 = 0$, Eq. c.

3. Transform the deformation equation to an axial-force equation

By substituting axial-force relations in Eq. c, the following equation is obtained: $10P_1(5)/(1.25E) - 16P_2(9)/(1.20E) + 6P_3(7.5)/E = 0$, or $40P_1 - 120P_2 + 45P_3 = 0$, Eq. c'.

4. Solve the simultaneous equations developed

Solve the simultaneous equations a, b, and c' to obtain $P_1 = 11,810$ lb (52,530 N); $P_2 = 5100$ lb (22,684 N); $P_3 = 3090$ lb (13,744 N).

5. Locate the center of rotation

To locate the center of rotation, compute the relative deformation of rods 1 and 2. Thus $\Delta l_1 = 11,810(5)/(1.25E) = 47,240/E$; $\Delta l_2 = 5100(9)/(1.20E) = 38,250/E$.

In Fig. 33c, by similar triangles, $x/(x-6) = \Delta l_1/\Delta l_2 = 1.235$; $x = 31.5$ ft (9.6 m).

6. Verify the computed values of the tensile forces

Calculate the moment with respect to A of the applied and resisting forces. Thus $M_{Aa} = 20,000(4) = 80,000$ lb·ft (108,400 N·m); $M_{Ar} = 5100(6) + 3090(16) = 80,000$ lb·ft (108,400 N·m). Since the moments are equal, the results are verified.

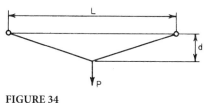

FIGURE 34

ANALYSIS OF CABLE SUPPORTING A CONCENTRATED LOAD

A cold-drawn steel wire ¼ in. (6.35 mm) in diameter is stretched tightly between two points lying on the same horizontal plane 80 ft (24.4 m) apart. The stress in the wire is 50,000 lb/sq.in. (344,700 kPa). A load of 200 lb (889.6 N) is suspended at the center of the cable. Determine the sag of the cable and the final stress in the cable. Verify that the results obtained are compatible.

Calculation Procedure:

1. Derive the stress and strain relations for the cable
With reference to Fig. 34, L = distance between supports, ft (m); P = load applied at center of cable span, lb (N); d = deflection of cable center, ft (m); ε = strain of cable caused by P; s_1 and s_2 = initial and final tensile stress in cable, respectively, lb/sq.in. (kPa).

Refer to the geometry of the deflection diagram. Taking into account that d/L is extremely small, derive the following approximations: $s_2 = PL/(4Ad)$, Eq. a; $\varepsilon = 2(d/L)^2$, Eq. b.

2. Relate stress and strain
Express the increase in stress caused by P in terms of ε, and apply the above two equations to derive $2E(d/L)^3 + s_1(d/L) = P/(4A)$, Eq. c.

3. Compute the deflection at the center of the cable
Using Eq. c, we get $2(30)(10)^6(d/L)^3 + 50,000 d/L = 200/[4(0.049)]$, so $d/L = 0.0157$ and $\therefore d = 0.0157(80) = 1.256$ ft (0.382 m).

4. Compute the final tensile stress
Write Eq. a as $s_2 = [P/(4A)]/(d/L) = 1020/0.0157 = 65,000$ lb/sq.in. (448,110 kPa).

5. Verify the results computed
To demonstrate that the results are compatible, accept the computed value of d/L as correct. Then apply Eq. b to find the strain, and compute the corresponding stress. Thus $\varepsilon = 2(0.0157)2 = 4.93 \times 10^{-4}$; $s_2 = s_1 + E\varepsilon = 50,000 + 30 \times 10^6 \times 4.93 \times 10^{-4} = 64,800$ lb/sq.in. (446,731 kPa). This agrees closely with the previously calculated stress of 65,000 lb/sq.in. (448,110 kPa).

DISPLACEMENT OF TRUSS JOINT

In Fig. 35a, the steel members AC and BC both have a cross-sectional area of 1.2 sq.in. (7.7 cm²). If a load of 20 kips (89.0 kN) is suspended at C, how much is joint C displaced?

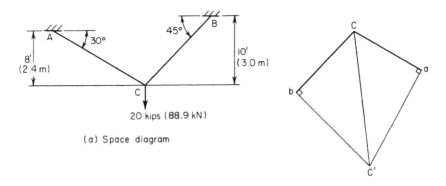

(a) Space diagram

(b) Displacement diagram

FIGURE 35

Calculation Procedure:

1. Compute the length of each member and the tensile forces
Consider joint C as a free body to find the tensile force in each member. Thus $L_{AC} = 192$ in. (487.7 cm); $L_{BC} = 169.7$ in. (431.0 cm); $PAC = 14{,}640$ lb (65,118.7 N); $P_{BC} = 17{,}930$ lb (79,752.6 N).

2. Determine the elongation of each member
Use the relation $\Delta l = PL/(AE)$. Thus $\Delta l_{AC} = 14{,}640(192)/[1.2(30 \times 10^6)] = 0.0781$ in. (1.983 mm); $\Delta l_{BC} = 17{,}930(169.7)/[1.2(30 \times 10^6)] = 0.0845$ in. (2.146 mm).

3. Construct the Williott displacement diagram
Selecting a suitable scale, construct the Williott displacement diagram as follows: Draw (Fig. 35b) line Ca parallel to member AC, with $Ca = 0.0781$ in. (1.98 mm). Similarly, draw Cb parallel to member BC, with $Cb = 0.0845$ in. (2.146 mm).

4. Determine the displacement
Erect perpendiculars to Ca and Cb at a and b, respectively. Designate the intersection point of these perpendiculars as C'.

Line CC' represents, in both magnitude and direction, the approximate displacement of joint C under the applied load. Scaling distance CC' to obtain the displacement shows that the displacement of $C = 0.134$ in. (3.4036 mm).

AXIAL STRESS CAUSED BY IMPACT LOAD

A body weighing 18 lb (80.1 N) falls 3 ft (0.9 m) before contacting the end of a vertical steel rod. The rod is 5 ft (1.5 m) long and has a cross-sectional area of 1.2 sq.in. (7.74 cm²). If the entire kinetic energy of the falling body is absorbed by the rod, determine the stress induced in the rod.

Calculation Procedure:

1. State the equation for the induced stress
Equate the energy imparted to the rod to the potential energy lost by the falling body: $s = (P/A)\{1 + [1 + 2Eh/(LP/A)]^{0.5}\}$, where h = vertical displacement of body, ft (m).

2. Substitute the numerical values
Thus, $P/A = 18/1.2 = 15$ lb/sq.in. (103 kPa); $h = 3$ ft (0.9 m); $L = 5$ ft (1.5 m); $[2Eh/(LP/A) = 2(30) \times (10^6)(3)]/[5(15)] = 2{,}400{,}000$. Then $s = 23{,}250$ lb/sq.in. (160,285.5 kPa).

Related Calculations. Where the deformation of the supporting member is negligible in relation to the distance h, as it is in the present instance, the following approximation is used: $s = [2PEh/(AL)]^{0.5}$.

STRESSES ON AN OBLIQUE PLANE

A prism $ABCD$ in Fig. 36a has the principal stresses of 6300- and 2400-lb/sq.in. (43,438.5- and 16,548.0-kPa) tension. Applying both the analytical and graphical methods, determine the normal and shearing stress on plane AE.

Calculation Procedure:

1. Compute the stresses, using the analytical method
A principal stress is a normal stress not accompanied by a shearing stress. The plane on which the principal stress exists is termed a *principal plane*. For a condition of plane stress, there are two principal planes through every point in a stressed body and these planes are mutually perpendicular. Moreover, one principal stress is the maximum normal stress existing at that point; the other is the minimum normal stress.

Let s_x and s_y = the principal stress in the x and y direction, respectively; s_n = normal stress on the plane making an angle θ with the y axis; s_s = shearing stress on this plane. All stresses are expressed in pounds per square inch (kilopascals) and all angles in degrees. Tensile stresses are positive; compressive stresses are negative.

Applying the usual stress equations yields $s_n = s_y + (s_x - s_y)\cos^2\theta$; $s_s = \frac{1}{2}(s_x - s_y)\sin 2\theta$. Substituting gives $s_n = 2400 + (6300 - 2400)0.766^2 = 4690$-lb/sq.in. (32,337.6-kPa) tension, and $s_s = \frac{1}{2}(6300 - 2400)0.985 = 1920$ lb/sq.in. (13,238.4 kPa).

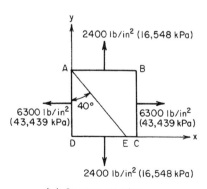

(a) Stresses on prism

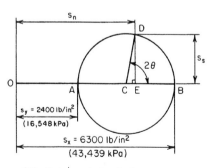

(b) Mohr's circle of stress

FIGURE 36

1.68 STRUCTURAL ENGINEERING

2. Apply the graphical method of solution
Construct, in Fig. 36b, Mohr's circle of stress thus: Using a suitable scale, draw $OA = s_y$, and $OB = s_x$. Draw a circle having AB as its diameter. Draw the radius CD making an angle of $2\theta = 80°$ with AB. Through D, drop a perpendicular DE to AB. Then $OE = s_n$ and $ED = s_s$. Scale OE and ED to obtain the normal and shearing stresses on plane AE.

Related Calculations. The normal stress may also be computed from $s_n = (s_x + s_y)0.5 + (s_x - s_y)0.5 \cos 2\theta$.

EVALUATION OF PRINCIPAL STRESSES

The prism $ABCD$ in Fig. 37a is subjected to the normal and shearing stresses shown. Construct Mohr's circle to determine the principal stresses at A, and locate the principal planes.

Calculation Procedure:

1. Draw the lines representing the normal stresses (Fig. 37b)
Through the origin O, draw a horizontal base line. Locate points E and F such that $OE = 8400$ lb/sq.in. (57,918.0 kPa) and $OF = 2000$ lb/sq.in. (13,790.0 kPa). Since both normal stresses are tensile, E and F lie to the right of O. Note that the construction required here is the converse of that required in the previous calculation procedure.

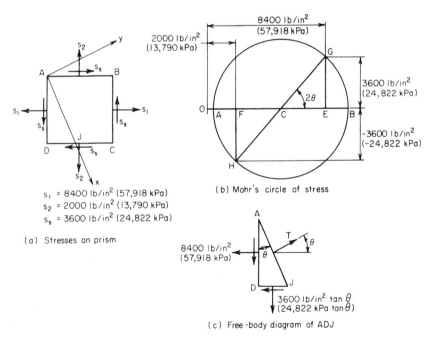

FIGURE 37

2. Draw the lines representing the shearing stresses
Construct the vertical lines EG and FH such that $EG = 3600$ lb/sq.in. (24,822.0 kPa), and $FH = -3600$ lb/sq.in. (−24,822.0 kPa).

3. Continue the construction
Draw line GH to intersect the base line at C.

4. Construct Mohr's circle
Draw a circle having GH as diameter, intersecting the base line at A and B. Then lines OA and OB represent the principal stresses.

5. Scale the diagram
Scale OA and OB to obtain $f_{max} = 10{,}020$ lb/sq.in. (69,087.9 kPa); $f_{min} = 380$ lb/sq.in. (2620.1 kPa). Both stresses are tension.

6. Determine the stress angle
Scale angle BCG and measure it as $48°22'$. The angle between the x axis, on which the maximum stress exists, and the side AD of the prism is one-half of BCG.

7. Construct the x and y axes
In Fig. 37a, draw the x axis, making a counterclockwise angle of $24°11'$ with AD. Draw the y axis perpendicular thereto.

8. Verify the locations of the principal planes
Consider ADJ as a free body. Set the length AD equal to unity. In Fig. 37c, since there is no shearing stress on AJ, $\Sigma F_H = T \cos \theta - 8400 - 3600 \tan \theta = 0$; $T \cos \theta = 8400 + 3600(0.45) = 10{,}020$ lb/sq.in. (69,087.9 kPa). The stress on $AJ = T/AJ = T \cos \theta = 10{,}020$ lb/sq.in. (69,087.9 kPa).

HOOP STRESS IN THIN-WALLED CYLINDER UNDER PRESSURE

A steel pipe 5 ft (1.5 m) in diameter and $3/8$ in. (9.53 mm) thick sustains a fluid pressure of 180 lb/sq.in. (1241.1 kPa). Determine the hoop stress, the longitudinal stress, and the increase in diameter of this pipe. Use 0.25 for Poisson's ratio.

Calculation Procedure:

1. Compute the hoop stress
Use the relation $s = pD/(2t)$, where $s =$ hoop or tangential stress, lb/sq.in. (kPa); $p =$ radial pressure, lb/sq.in. (kPa); $D =$ internal diameter of cylinder, in (mm); $t =$ cylinder wall thickness, in. (mm). Thus, for this cylinder, $s = 180(60)/[2(3/8)] = 14{,}400$ lb/sq.in. (99,288.0 kPa).

2. Compute the longitudinal stress
Use the relation $s' = pD/(4t)$, where $s' =$ longitudinal stress, i.e., the stress parallel to the longitudinal axis of the cylinder, lb/sq.in. (kPa), with other symbols as before. Substituting yields $s' = 7200$ lb/sq.in. (49,644.0 kPa).

3. Compute the increase in the cylinder diameter
Use the relation $\Delta D = (D/E)(s - vs')$, where $v =$ Poisson's ratio. Thus $\Delta D = 60(14{,}400 - 0.25 \times 7200)/(30 \times 10^6) = 0.0252$ in. (0.6401 mm).

STRESSES IN PRESTRESSED CYLINDER

A steel ring having an internal diameter of 8.99 in. (228.346 mm) and a thickness of ¼ in. (6.35 mm) is heated and allowed to shrink over an aluminum cylinder having an external diameter of 9.00 in. (228.6 mm) and a thickness of ½ in. (12.7 mm). After the steel cools, the cylinder is subjected to an internal pressure of 800 lb/sq.in. (5516 kPa). Find the stresses in the two materials. For aluminum, $E = 10 \times 10^6$ lb/sq.in. (6.895 × 10^7 kPa).

Calculation Procedure:

1. Compute the radial pressure caused by prestressing
Use the relation $p = 2\Delta D/\{D^2[1/(t_a E_a) + 1/(t_s E_s)]\}$, where p = radial pressure resulting from prestressing, lb/sq.in. (kPa), with other symbols the same as in the previous calculation procedure and the subscripts a and s referring to aluminum and steel, respectively. Thus, $p = 2(0.01)/\{9^2[1/(0.5 \times 10 \times 10^6) + 1/(0.25 \times 30 \times 10^6)]\} = 741$ lb/sq.in. (5109.2 kPa).

2. Compute the corresponding prestresses
Using the subscripts 1 and 2 to denote the stresses caused by prestressing and internal pressure, respectively, we find $s_{a1} = pD/(2t_a)$, where the symbols are the same as in the previous calculation procedure. Thus, $s_{a1} = 741(9)/[2(0.5)] = 6670$-lb/sq.in. (45,989.7-kPa) compression. Likewise, $s_{s1} = 741(9)/[2(0.25)] = 13,340$-lb/sq.in. (91,979-kPa) tension.

3. Compute the stresses caused by internal pressure
Use the relation $s_{s2}/s_{a2} = E_s/E_a$ or, for this cylinder, $s_{s2}/s_{a2} = (30 \times 10^6)/(10 \times 10^6) = 3$. Next, compute s_{a2} from $t_a s_{a2} t_s s_{s2} = pD/2$, or $s_{a2} = 800(9)/[2(0.5 + 0.25 \times 3)] = 2880$-lb/sq.in. (19,857.6-kPa) tension. Also, $s_{s2} = 3(2880) = 8640$-lb/sq.in. (59,572.8-kPa) tension.

4. Compute the final stresses
Sum the results in steps 2 and 3 to obtain the final stresses: $s_{a3} = 6670 - 2880 = 3790$-lb/sq.in. (26,132.1-kPa) compression; $s_{s3} = 13,340 + 8640 = 21,980$-lb/sq.in. (151,552.1-kPa) tension.

5. Check the accuracy of the results
Ascertain whether the final diameters of the steel ring and aluminum cylinder are equal. Thus, setting $s' = 0$ in. $\Delta D = (D/E)(s - vs')$, we find $\Delta D_a = -3790(9)/(10 \times 10^6) = -0.0034$ in. (−0.0864 mm), $D_a = 9.0000 - 0.0034 = 8.9966$ in. (228.51 mm). Likewise, $\Delta D_s = 21,980(9)/(30 \times 10^6) = 0.0066$ in. (0.1676 mm), $D_s = 8.99 + 0.0066 = 8.9966$ in. (228.51 mm). Since the computed diameters are equal, the results are valid.

HOOP STRESS IN THICK-WALLED CYLINDER

A cylinder having an internal diameter of 20 in. (508 mm) and an external diameter of 36 in. (914 mm) is subjected to an internal pressure of 10,000 lb/sq.in. (68,950 kPa) and an external pressure of 2000 lb/sq.in. (13,790 kPa) as shown in Fig. 38. Determine the hoop stress at the inner and outer surfaces of the cylinder.

Calculation Procedure:

1. Compute the hoop stress at the inner surface of the cylinder
Use the relation $s_i = [p_1(r_1^2 + r_2^2) - 2p_2 r_2^2]/(r_2^2 - r_1^2)$, where s_i = hoop stress at inner surface, lb/sq.in. (kPa); p_1 = internal pressure, lb/sq.in. (kPa); r_1 = internal radius, in. (mm); r_2 = external

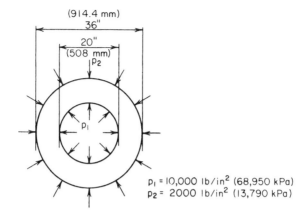

FIGURE 38. Thick-walled cylinder under internal and external pressure.

radius, in. (mm); p_2 = external pressure, lb/sq.in. (kPa). Substituting gives $s_i = [10,000(100 + 324) - 2(2000)(324)]/(324 - 100) = 13,100$-lb/sq.in. (90,324.5-kPa) tension.

2. Compute the hoop stress at the outer cylinder surface
Use the relation $s_o = [2p_1 r_1^2 - p_2(r_1^2 + r_2^2)]/(r_2^2 - r_1^2)$, where the symbols are as before. Substituting gives $s_o = [2(10,000)(100) - 2000(100 + 324)]/(324 - 100) = 5100$-lb/sq.in. (35,164.5-kPa) tension.

3. Check the accuracy of the results
Use the relation $s_1 r_1 - s_0 r_2 = [(r_2 - r_1)/(r_2 + r_1)](p_1 r_1 + p_2 r_2)$. Substituting the known values verifies the earlier calculations.

THERMAL STRESS RESULTING FROM HEATING A MEMBER

A steel member 18 ft (5.5 m) long is set snugly between two walls and heated 80°F (44.4°C). If each wall yields 0.015 in. (0.381 mm), what is the compressive stress in the member? Use a coefficient of thermal expansion of 6.5×10^{-6}/°F (1.17×10^{-5}/°C) for steel.

Calculation Procedure:

1. Compute the thermal expansion of the member without restraint
Replace the true condition of partial restraint with the following equivalent conditions: The member is first allowed to expand freely under the temperature rise and is then compressed to its true final length.

To compute the thermal expansion without restraint, use the relation $\Delta L = cL\Delta T$, where c = coefficient of thermal expansion, /°F (/°C); ΔT = increase in temperature, °F (°C); L = original length of member, in. (mm); ΔL = increase in length of the member, in. (mm). Substituting gives $\Delta L = 6.5(10^{-6})(18)(12)(80) = 0.1123$ in. (2.852 mm).

2. Compute the linear restraint exerted by the walls
The walls yield 2(0.015) = 0.030 in. (0.762 mm). Thus, the restraint exerted by the walls is $\Delta L_w = 0.1123 - 0.030 = 0.0823$ in. (2.090 mm).

3. Determine the compressive stress
Use the relation $s = E\Delta L/L$, where the symbols are as given earlier. Thus, $s = 30(10^6)(0.0823)/[18(12)] = 11{,}430$ lb/sq.in. (78,809.9 kPa).

THERMAL EFFECTS IN COMPOSITE MEMBER HAVING ELEMENTS IN PARALLEL

A ½-in.-diameter (12.7-mm) Copperweld bar consists of a steel core ⅜ in. (9.53 mm) in diameter and a copper skin 1/16 in. (1.6 mm) thick. What is the elongation of a 1-ft (0.3-m) length of this bar, and what is the internal force between the steel and copper arising from a temperature rise of 80°F (44.4°C)? Use the following values for thermal expansion coefficients: $c_s = 6.5 \times 10^{-6}$ and $c_c = 9.0 \times 10^{-6}$, where the subscripts s and c refer to steel and copper, respectively. Also, $E_c = 15 \times 10^6$ lb/sq.in. (1.03 × 10^8 kPa).

Calculation Procedure:

1. Determine the cross-sectional areas of the metals
The total area $A = 0.1963$ sq.in. (1.266 cm^2). The area of the steel $A_s = 0.1105$ sq.in. (0.712 cm^2). By difference, the area of the copper $A_c = 0.0858$ sq.in. (0.553 cm^2).

2. Determine the coefficient of expansion of the composite member
Weight the coefficients of expansion of the two members according to their respective AE values. Thus

$A_s E_s$ (relative) = 0.1105 × 30 × 10^6 = 3315
$A_c E_c$ (relative) = 0.0858 × 15 × 10^6 = 1287
Total   4602

Then the coefficient of thermal expansion of the composite member is $c = (3315c_s + 1287c_c)/4602 = 7.2 \times 10^{-6}/°F$ (1.30 × 10^{-5}/°C).

3. Determine the thermal expansion of the 1-ft (0.3-m) section
Using the relation $\Delta L = cL\Delta T$, we get $\Delta L = 7.2(10^6)(12)(80) = 0.00691$ in. (0.17551 mm).

4. Determine the expansion of the first material without restraint
Using the same relation as in step 3 for copper *without* restraint yields $\Delta L_c = 9.0(10^{-6}) \times (12)(80) = 0.00864$ in. (0.219456 mm).

5. Compute the restraint of the first material
The copper is restrained to the amount computed in step 3. Thus, the restraint exerted by the steel is $\Delta L_{cs} = 0.00864 - 0.00691 = 0.00173$ in. (0.043942 mm).

6. Compute the restraining force exerted by the second material
Use the relation $P = (A_c E_c \Delta L_{cs})/L$, where the symbols are as given before: $P = [1{,}287{,}000(0.00173)]/12 = 185$ lb (822.9 N).

7. Verify the results obtained
Repeat steps 4, 5, and 6 with the two materials interchanged. So $\Delta L_s = 6.5(10^{-6})(12)(80) = 0.00624$ in. (0.15849 mm); $\Delta L_{sc} = 0.00691 - 0.00624 = 0.00067$ in. (0.01701 mm). Then $P = 3,315,000(0.00067)/12 = 185$ lb (822.9 N), as before.

THERMAL EFFECTS IN COMPOSITE MEMBER HAVING ELEMENTS IN SERIES

The aluminum and steel bars in Fig. 39 have cross-sectional areas of 1.2 and 1.0 sq.in. (7.7 and 6.5 cm²), respectively. The member is restrained against lateral deflection. A temperature rise of 100°F (55°C) causes the length of the member to increase to 42.016 in. (106.720 cm). Determine the stress and deformation of each bar. For aluminum, $E = 10 \times 10^6$ $c = 13.0 \times 10^{-6}$; for steel, $c = 6.5 \times 10^{-6}$.

Calculation Procedure:

1. Express the deformation of each bar resulting from the temperature change and the compressive force
The temperature rise causes the bar to expand, whereas the compressive force resists this expansion. Thus, the net expansion is the difference between these two changes, or $\Delta L_a = c L \Delta T - PL/(AE)$, where the subscript a refers to the aluminum bar; the other symbols are the same as given earlier. Substituting gives $\Delta L_a = 13.0 \times 10^{-6}(24)(100) - P(24)/[1.2(10 \times 10^6)] = (31,200 - 2P)10^{-6}$, Eq. a. Likewise, for steel: $\Delta L_s = 6.5 \times 10^{-6}(18)(100) - P(18)/[1.0(30 \times 10^6)] = (11,700 - 0.6P)10^{-6}$, Eq. b.

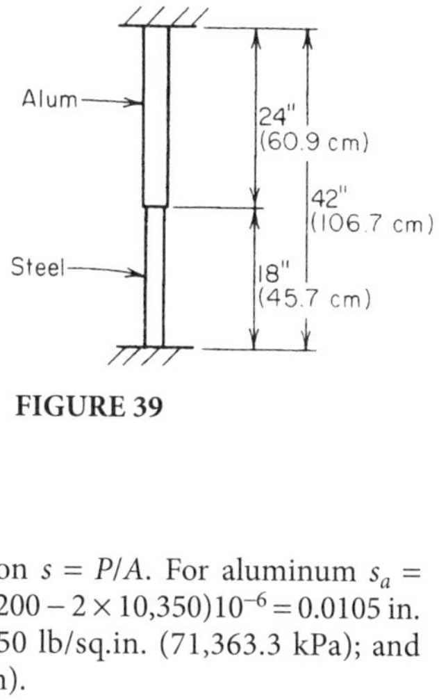

2. Sum the results in step 1 to obtain the total deformation of the member
Set the result equal to 0.016 in. (0.4064 mm); solve for P. Or, $\Delta L = (42,900 - 2.6P)10^{-6} = 0.016$ in. (0.4064 mm); $P = (42,900 - 16,000)/2.6 = 10,350$ lb (46,037 N).

FIGURE 39

3. Determine the stresses and deformation
Substitute the computed value of P in the stress equation $s = P/A$. For aluminum $s_a = 10,350/1.2 = 8630$ lb/sq.in. (59,503.9 kPa). Then $\Delta L_a = (31,200 - 2 \times 10,350)10^{-6} = 0.0105$ in. (0.2667 mm). Likewise, for steel $s_s = 10,350/1.0 = 10,350$ lb/sq.in. (71,363.3 kPa); and $\Delta L_s = (11,700 - 0.6 \times 10,350)10^{-6} = 0.0055$ in. (0.1397 mm).

SHRINK-FIT STRESS AND RADIAL PRESSURE

An open steel cylinder having an internal diameter of 4 ft (1.2 m) and a wall thickness of 5/16 in. (7.9 mm) is to be heated to fit over an iron casting. The internal diameter of the cylinder before heating is 1/32 in. (0.8 mm) less than that of the casting. How much must the temperature of the cylinder be increased to provide a clearance of 1/32 in. (0.8 mm) all around between the cylinder and casting? If the casting is considered rigid, what stress will exist in the cylinder after it cools, and what radial pressure will it then exert on the casting?

Calculation Procedure:

1. Compute the temperature rise required
Use the relation $\Delta T = \Delta D/(cD)$, where ΔT = temperature rise required, °F (°C); ΔD = change in cylinder diameter, in. (mm); c = coefficient of expansion of the cylinder = 6.5×10^{-6}/°F (1.17×10^{-5}/°C); D = cylinder internal diameter before heating, in. (mm). Thus $\Delta T = (3/32)/[6.5 \times 10^{-6}(48)] = 300$°F (167°C).

2. Compute the hoop stress in the cylinder
Upon cooling, the cylinder has a diameter $1/32$ in. (0.8 mm) larger than originally. Compute the hoop stress from $s = E\Delta D/D = 30 \times 10^6(1/32)/48 = 19{,}500$ lb/sq.in. (134,452.5 kPa).

3. Compute the associated radial pressure
Use the relation $p = 2ts/D$, where p = radial pressure, lb/sq.in. (kPa), with the other symbols as given earlier. Thus $p = 2(5/16)(19{,}500)/48 = 254$ lb/sq.in. (1751.3 kPa).

TORSION OF A CYLINDRICAL SHAFT

A torque of 8000 lb·ft (10,840 N·m) is applied at the ends of a 14-ft (4.3-m) long cylindrical shaft having an external diameter of 5 in. (127 mm) and an internal diameter of 3 in. (76.2 mm). What are the maximum shearing stress and the angle of twist of the shaft if the modulus of rigidity of the shaft is 6×10^6 lb/sq.in. (4.1×10^4 MPa)?

Calculation Procedure:

1. Compute the polar moment of inertia of the shaft
For a hollow circular shaft, $J = (\pi/32)(D^4 - d^4)$, where J = polar moment of inertia of a transverse section of the shaft with respect to the longitudinal axis, in^4 (cm^4); D = external diameter of shaft, in. (mm); d = internal diameter of shaft, in. (mm). Substituting gives $J = (\pi/32)(5^4 - 3^4) = 53.4$ in^4 (2222.6 cm^4).

2. Compute the shearing stress in the shaft
Use the relation $s_s = TR/J$, where s_s = shearing stress, lb/sq.in. (MPa); T = applied torque, lb·in. (N·m); H = radius of shaft, in. (mm). Thus $s_s = [(8000)(12)(2.5)]/53.4 = 4500$ lb/sq.in. (31,027.5 kPa).

3. Compute the angle of twist of the shaft
Use the relation $\theta = TL/JG$, where θ = angle of twist, rad; L = shaft length, in. (mm); G = modulus of rigidity, lb/sq.in. (GPa). Thus $\theta = (8000)(12)(14)(12)/[53.4(6{,}000{,}000)] = 0.050$ rad, or 2.9°.

ANALYSIS OF A COMPOUND SHAFT

The compound shaft in Fig. 40 was formed by rigidly joining two solid segments. What torque may be applied at B if the shearing stress is not to exceed 15,000 lb/sq.in. (103.4 MPa) in the steel and 10,000 lb/sq.in. (69.0 MPa) in the bronze? Here $G_s = 12 \times 10^6$ lb/sq.in. (82.7 GPa); $G_b = 6 \times 10^6$ lb/sq.in. (41.4 GPa).

Calculation Procedure:

1. Determine the relationship between the torque in the shaft segments

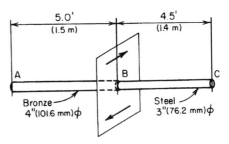

FIGURE 40. Compound shaft.

Since segments AB and BC (Fig. 40) are twisted through the same angle, the torque applied at the junction of these segments is distributed in proportion to their relative rigidities. Using the subscripts s and b to denote steel and bronze, respectively, we see that $\theta = T_s L_s/(J_s G_s) = T_b L_b/(J_b G_b)$, where the symbols are as given in the previous calculation procedure. Solving yields $T_s = (5/4.5)(34/44)(12/6)T_b = 0.703\,T_b$.

2. Establish the relationship between the shearing stresses

For steel, $s_{ss} = 16T_s/(\pi D^3)$, where the symbols are as given earlier. Thus $s_{ss} = 16(0.703T_b)/(\pi 3^3)$. Likewise, for bronze, $s_{sb} = 16T_b/(\pi 4^3)$, ∴ $s_{ss} = 0.703(4^3/3^3)s_{sb} = 1.67s_{sb}$.

3. Compute the allowable torque

Ascertain which material limits the capacity of the member, and compute the allowable torque by solving the shearing-stress equation for T.

If the bronze were stressed to 10,000 lb/sq.in. (69.0 MPa), inspection of the above relations shows that the steel would be stressed to 16,700 lb/sq.in. (115.1 MPa), which exceeds the allowed 15,000 lb/sq.in. (103.4 MPa). Hence, the steel limits the capacity. Substituting the allowed shearing stress of 15,000 lb/sq.in. (103.4 MPa) gives $T_s = 15,000\pi(3^3)/[16(12)] = 6630$ lb·ft (8984.0 N·m); also, $T_b = 6630/0.703 = 9430$ lb·ft (12,777.6 N·m). Then $T = 6630 + 9430 = 16,060$ lb·ft (21,761.3 N·m).

Stresses in Flexural Members

In the analysis of beam action, the general assumption is that the beam is in a horizontal position and carries vertical loads lying in an axis of symmetry of the transverse section of the beam.

The vertical shear V at a given section of the beam is the algebraic sum of all vertical forces to the left of the section, with an upward force being considered positive.

The bending moment M at a given section of the beam is the algebraic sum of the moments of all forces to the left of the section with respect to that section, a clockwise moment being considered positive.

If the proportional limit of the beam material is not exceeded, the bending stress (also called the flexural, or fiber, stress) at a section varies linearly across the depth of the section, being zero at the neutral axis. A positive bending moment induces compressive stresses in the fibers above the neutral axis and tensile stresses in the fibers below. Consequently, the elastic curve of the beam is concave upward where the bending moment is positive.

SHEAR AND BENDING MOMENT IN A BEAM

Construct the shear and bending-moment diagrams for the beam in Fig. 41. Indicate the value of the shear and bending moment at all significant sections.

Calculation Procedure:

1. Replace the distributed load on each interval with its equivalent concentrated load

Where the load is uniformly distributed, this equivalent load acts at the center of the interval of the beam. Thus $W_{AB} = 2(4) = 8$ kips (35.6 kN); $W_{BC} = 2(6) = 12$ kips (53.3 kN); $W_{AC} = 8 + 12 = 20$ kips (89.0 kN); $W_{CD} = 3(15) = 45$ kips (200.1 kN); $W_{DE} = 1.4(5) = 7$ kips (31.1 kN).

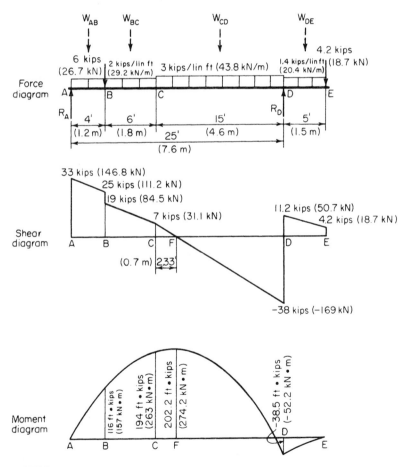

FIGURE 41

2. Determine the reaction at each support

Take moments with respect to the other support. Thus $\Sigma M_D = 25R_A - 6(21) - 20(20) - 45(7.5) + 7(2.5) + 4.2(5) = 0$; $\Sigma M_A = 6(4) + 20(5) + 45(17.5) + 7(27.5) + 4.2(30) - 25R_D = 0$. Solving gives $R_A = 33$ kips (146.8 kN); $R_D = 49.2$ kips (218.84 kN).

3. Verify the computed results and determine the shears

Ascertain that the algebraic sum of the vertical forces is zero. If this is so, the computed results are correct.

Starting at A, determine the shear at every significant section, or directly to the left or right of that section if a concentrated load is present. Thus V_A at right = 33 kips (146.8 kN); V_B at left = $33 - 8 = 25$ kips (111.2 kN); V_B at right = $25 - 6 = 19$ kips (84.5 kN); $V_C = 19 - 12 = 7$ kips (31.1 kN); V_D at left = $7 - 45 = -38$ kips (-169.0 kN); V_D at right = $-38 + 49.2 = 11.2$ kips (49.8 kN); V_E at left = $11.2 - 7 = 4.2$ kips (18.7 kN); V_E at right = $4.2 - 4.2 = 0$.

4. Plot the shear diagram

Plot the points representing the forces in the previous step in the shear diagram. Since the loading between the significant sections is uniform, connect these points with straight lines. In general, the slope of the shear diagram is given by $dV/dx = -w$, where w = unit load at the given section and x = distance from left end to the given section.

5. Determine the bending moment at every significant section

Starting at A, determine the bending moment at every significant section. Thus $M_A = 0$; $M_B = 33(4) - 8(2) = 116$ ft·kips (157 kN·m); $M_C = 33(10) - 8(8) - 6(6) - 12(3) = 194$ ft·kips (263 kN·m). Similarly, $M_D = -38.5$ ft·kips (-52.2 kN·m); $M_E = 0$.

6. Plot the bending-moment diagram

Plot the points representing the values in step 5 in the bending-moment diagram (Fig. 41). Complete the diagram by applying the slope equation $dM/dx = V$, where V denotes the shear at the given section. Since this shear varies linearly between significant sections, the bending-moment diagram comprises a series of parabolic arcs.

7. Alternatively, apply a moment theorem

Use this theorem: If there are no externally applied moments in an interval 1-2 of the span, the difference between the bending moments is $M_2 - = M_1 = \int_1^2 V\, dx$ = the area under the shear diagram across the interval.

Calculate the areas under the shear diagram to obtain the following results: $M_A = 0$; $M_B = M_A + \frac{1}{2}(4)(33 + 25) = 116$ ft·kips (157.3 kN·m); $M_C = 116 + \frac{1}{2}(6)(19 + 7) = 194$ ft·kips (263 kN·m); $M_D = 194 + \frac{1}{2}(15)(7 - 38) = -38.5$ ft·kips (-52.2 kN·m); $M_E = -38.5 + \frac{1}{2}(5)(11.2 + 4.2) = 0$.

8. Locate the section at which the bending moment is maximum

As a corollary of the equation in step 6, the maximum moment occurs where the shear is zero or passes through zero under a concentrated load. Therefore, $CF = 7/3 = 2.33$ ft (0.710 m).

9. Compute the maximum moment

Using the computed value for CF, we find $M_F = 194 + \frac{1}{2}(2.33)(7) = 202.2$ ft·kips (274.18 kN·m).

BEAM BENDING STRESSES

A beam having the trapezoidal cross section shown in Fig. 42a carries the loads indicated in Fig. 42b. What is the maximum bending stress at the top and at the bottom of this beam?

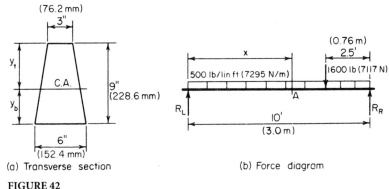

FIGURE 42

Calculation Procedure:

1. Compute the left reaction and the section at which the shear is zero

The left reaction $R_L = \frac{1}{2}(10)(500) + 1600(2.5/10) = 2900$ lb (12,899.2 N). The section A at which the shear is zero is $x = 2900/500 = 5.8$ ft (1.77 m).

2. Compute the maximum moment

Use the relation $M_A = \frac{1}{2}(2900)(5.8) = 8410$ lb·ft (11,395.6 N·m) = 100,900 lb·in. (11,399.682 N·m).

3. Locate the centroidal axis of the section

Use the AISC *Manual* for properties of the trapezoid. Or $y_t = (9/3)[(2 \times 6 + 3)1(6 + 3)] = 5$ in. (127 mm); $y_b = 4$ in. (101.6 mm).

4. Compute the moment of inertia of the section

Using the AISC *Manual*, $I = (9^3/36)[(6^2 + 4 \times 6 \times 3 + 3^2)/(6 + 3)] = 263.3$ in^4 (10,959.36 cm^4).

5. Compute the stresses in the beam

Use the relation $f = My/I$, where f = bending stress in a given fiber, lb/sq.in. (kPa); y = distance from neutral axis to given fiber, in. Thus $f_{top} = 100,900(5)/263.3 = 1916$-lb/sq.in. (13,210.8-kPa) compression, $f_{bottom} = 100,900(4)/263.3 = 1533$-lb/sq.in. (10,570.0-kPa) tension.

In general, the maximum bending stress at a section where the moment is M is given by $f = Mc/I$, where c = distance from the neutral axis to the outermost fiber, in. (mm). For a section that is symmetric about its centroidal axis, it is convenient to use the section modulus S of the section, this being defined as $S = I/c$. Then $f = M/S$.

ANALYSIS OF A BEAM ON MOVABLE SUPPORTS

The beam in Fig. 43a rests on two movable supports. It carries a uniform live load of w lb/lin ft and a uniform dead load of $0.2w$ lb/lin ft. If the allowable bending stresses in tension and compression are identical, determine the optimal location of the supports.

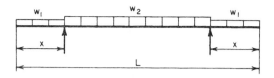

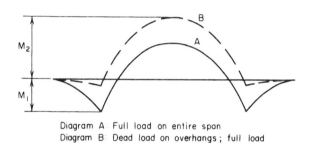

(a) Loads carried by overhanging beam

(b) Bending-moment diagrams

FIGURE 43

Calculation Procedure:

1. Place full load on the overhangs, and compute the negative moment

Refer to the moment diagrams. For every position of the supports, there is a corresponding maximum bending stress. The position for which this stress has the smallest value must be identified.

As the supports are moved toward the interior of the beam, the bending moments between the supports diminish in algebraic value. The optimal position of the supports is that for which the maximum potential negative moment M_1 is numerically equal to the maximum potential positive moment M_2. Thus, $M_1 = -1.2w(x^2/2) = -0.6wx^2$.

2. Place only the dead load on the overhangs and the full load between the supports. Compute the positive moment

Sum the areas under the shear diagram to compute M_2. Thus, $M_2 = \frac{1}{2}[1.2w(L/2 - x)^2 - 0.2wx^2] = w(0.15L^2 - 0.6Lx + 0.5x^2)$.

3. Equate the absolute values of M_1 and M_2 and solve for x

Substituting gives $0.6x^2 = 0.15L^2 - 0.6Lx + 0.5x^2$; $x = L\overline{10.5}^{0.5} - 3) = 0.240L$.

FLEXURAL CAPACITY OF A COMPOUND BEAM

A W16 × 45 steel beam in an existing structure was reinforced by welding a WT6 × 20 to the bottom flange, as in Fig. 44. If the allowable bending stress is 20,000 lb/sq.in. (137,900 kPa), determine the flexural capacity of the built-up member.

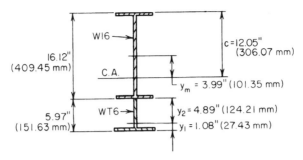

FIGURE 44. Compound beam.

Calculation Procedure:

1. Obtain the properties of the elements
Using the AISC *Manual*, determine the following properties. For the W16 × 45, $d = 16.12$ in. (409.45 mm); $A = 13.24$ sq.in. (85.424 cm^2); $I = 583$ in^4 (24,266 cm^4). For the WT6 × 20, $d = 5.97$ in. (151.63 mm); $A = 5.89$ sq.in. (38.002 cm^2); $I = 14$ in^4 (582.7 cm^4); $y_1 = 1.08$ in. (27.43 mm); $y_2 = 5.97 - 1.08 = 4.89$ in. (124.21 mm).

2. Locate the centroidal axis of the section
Locate the centroidal axis of the section with respect to the centerline of the W16 × 45, and compute the distance c from the centroidal axis to the outermost fiber. Thus, $y_m = 5.89[(8.06 + 4.89)]/(5.89 + 13.24) = 3.99$ in. (101.346 mm). Then $c = 8.06 + 3.99 = 12.05$ in. (306.07 mm).

3. Find the moment of inertia of the section with respect to its centroidal axis
Use the relation $I_0 + Ak^2$ for each member, and take the sum for the two members to find I for the built-up beam. Thus, for the W16 × 45: $k = 3.99$ in. (101.346 mm); $10 + Ak^2$ 583 + 13.24(3.99)^2 = 793 in^4 (33,007.1 cm^4). For the WT6 × 20: $k = 8.06 - 3.99 + 4.89 = 8.96$ in. (227.584 mm); $I_0 + Ak^2 = 14 + 5.89(8.96)^2 = 487$ in^4 (20,270.4 cm^4). Then $I = 793 + 487 = 1280$ in^4 (53,277.5 cm^4).

4. Apply the moment equation to find the flexural capacity
Use the relation $M = fI/c = 20{,}000(1280)/[12.05(12)] = 177{,}000$ lb·ft (240,012 N·m).

ANALYSIS OF A COMPOSITE BEAM

An 8 × 12 in. (203.2 × 304.8 mm) timber beam (exact size) is reinforced by the addition of a 7 × ½ in. (177.8 × 12.7 mm) steel plate at the top and a 7-in. (177.8-mm) 9.8-lb (43.59-N) steel channel at the bottom, as shown in Fig. 45a. The allowable bending stresses are 22,000 lb/sq.in. (151,690 kPa) for steel and 1200 lb/sq.in. (8274 kPa) for timber. The modulus of elasticity of the timber is 1.2×10^6 lb/sq.in. (8.274×10^6 kPa). How does the flexural strength of the reinforced beam compare with that of the original timber beam?

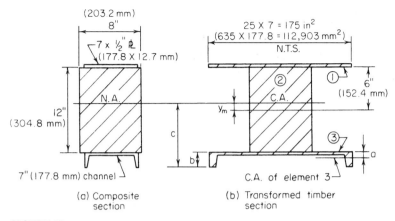

FIGURE 45

Calculation Procedure:

1. Compute the rigidity of the steel compared with that of the timber
Let n = the relative rigidity of the steel and timber. Then $n = E_s/E_t = (30 \times 10^6)/(1.2 \times 10^6) = 25$.

2. Transform the composite beam to an equivalent homogeneous beam
To accomplish this transformation, replace the steel with timber. Sketch the cross section of the transformed beam as in Fig. 45b. Determine the sizes of the hypothetical elements by retaining the dimensions normal to the axis of bending but multiplying the dimensions parallel to this axis by n.

3. Record the properties of each element of the transformed section
Element 1: $A = 25(7)(\frac{1}{2}) = 87.5$ sq.in. (564.55 cm^2); I_0 is negligible.
Element 2: $A = 8(12) = 96$ sq.in. (619.4 cm^2); $I_0 = \frac{1}{2}(8)12^3 = 1152$ in^4 (4.795 dm^4).
Element 3: Refer to the AISC *Manual* for the data; $A = 25(2.85) = 71.25$ sq.in. (459.71 cm^2); $I_0 = 25(0.98) = 25$ in^4 (1040.6 cm^4); $a = 0.55$ in. (13.97 mm); $b = 2.09$ in. (53.09 mm).

4. Locate the centroidal axis of the transformed section
Take static moments of the areas with respect to the centerline of the 8 × 12 in. (203.2 × 304.8 mm) rectangle. Then $y_m = [87.5(6.25) - 71.25(6.55)]/(87.5 + 96 + 71.25) = 0.31$ in. (7.87 mm). The neutral axis of the composite section is at the same location as the centroidal axis of the transformed section.

5. Compute the moment of inertia of the transformed section
Apply the relation in step 3 of the previous calculation procedure. Then compute the distance c to the outermost fiber. Thus, $I = 1152 + 25 + 87.5(6.25 - 0.31)^2 + 96(0.31)^2 + 71.25(6.55 + 0.31)^2 = 7626$ in^4 (31.74 dm^4). Also, $c = 0.31 + 6 + 2.09 = 8.40$ in. (213.36 mm).

6. Determine which material limits the beam capacity
Assume that the steel is stressed to capacity, and compute the corresponding stress in the transformed beam. Thus, $f = 22,000/25 = 880$ lb/sq.in. (6067.6 kPa) < 1200 lb/sq.in. (8274 kPa).

In the actual beam, the maximum timber stress, which occurs at the back of the channel, is even less than 880 lb/sq.in. (6067.6 kPa). Therefore, the strength of the member is controlled by the allowable stress in the steel.

7. Compare the capacity of the original and reinforced beams
Let subscripts 1 and 2 denote the original and reinforced beams, respectively. Compute the capacity of these members, and compare the results. Thus $M_1 = fI/c = 1200(1152)/6 = 230,000$ lb·in. (25,985.4 N·m); $M_2 = 880(7626)/8.40 = 799,000$ lb·in. (90,271.02 N·m); $M_2/M_1 = 799,000/230,000 = 3.47$. Thus, the reinforced beam is nearly 3½ times as strong as the original beam, before reinforcing.

BEAM SHEAR FLOW AND SHEARING STRESS

A timber beam is formed by securely bolting a 3 × 6 in. (76.2 × 152.4 mm) member to a 6 × 8 in. (152.4 × 203.2 mm) member (exact size), as shown in Fig. 46. If the beam carries a uniform load of 600 lb/lin ft (8.756 kN/m) on a simple span of 13 ft (3.9 m), determine the longitudinal shear flow and the shearing stress at the juncture of the two elements at a section 3 ft (0.91 m) from the support.

Calculation Procedure:

1. Compute the vertical shear at the given section
Shear flow is the shearing force acting on a unit distance. In this instance, the shearing force on an area having the same width as the beam and a length of 1 in. (25.4 mm) measured along the beam span is required.

Using dimensions and data from Fig. 46, we find $R = \frac{1}{2}(600)(13) = 3900$ lb (17,347.2 N); $V = 3900 - 3(600) = 2100$ lb (9340.8 N).

2. Compute the moment of inertia of the cross section

$$I = (^1/_{12})(bd^3) = (^1/_{12})(6)(11)^3 = 666 \text{ in}^4 \text{ (2.772 dm}^4\text{)}$$

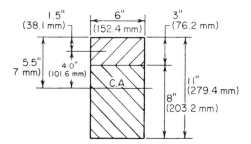

FIGURE 46

3. Determine the static moment of the cross-sectional area
Calculate the static moment Q of the cross-sectional area above the plane under consideration with respect to the centroidal axis of the section. Thus, $Q = Ay = 3(6)(4) = 72$ in^3 (1180.1 cm^3).

4. Compute the shear flow
Compute the shear flow q, using $q = VQ/I = 2100(72)/666 = 227$ lb/lin in. (39.75 kN/m).

5. Compute the shearing stress
Use the relation $v = q/t = VQ/(It)$, where $t =$ width of the cross section at the given plane. Then $v = 227/6 = 38$ lb/sq.in. (262.0 kPa).

Note that v represents both the longitudinal and the transverse shearing stress at a particular point. This is based on the principle that the shearing stresses at a given point in two mutually perpendicular directions are equal.

LOCATING THE SHEAR CENTER OF A SECTION

A cantilever beam carries the load shown in Fig. 47a and has the transverse section shown in Fig. 47b. Locate the shear center of the section.

Calculation Procedure:

1. Construct a free-body diagram of a portion of the beam
Consider that the transverse section of a beam is symmetric solely about its horizontal centroidal axis. If bending of the beam is not to be accompanied by torsion, the vertical

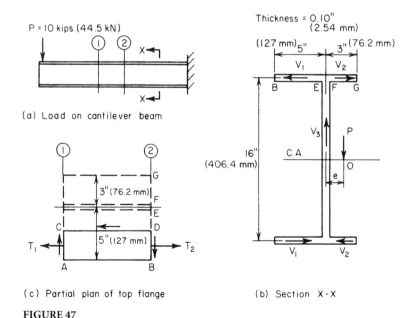

FIGURE 47

shearing force at any section must pass through a particular point on the centroidal axis designated as the *shear*, or *flexural*, center.

Cut the beam at section 2, and consider the left portion of the beam as a free body. In Fig. 47b, indicate the resisting shearing forces V_1, V_2, and V_3 that the right-hand portion of the beam exerts on the left-hand portion at section 2. Obtain the directions of V_1 and V_2 this way: Isolate the segment of the beam contained between sections 1 and 2; then isolate a segment *ABDC* of the top flange, as shown in Fig. 47c. Since the bending stresses at section 2 exceed those at section 1, the resultant tensile force T_2 exceeds T_1. The resisting force on *CD* is therefore directed to the left. From the equation of equilibrium $\Sigma M = 0$ it follows that the resisting shears on *AC* and *BD* have the indicated direction to constitute a clockwise couple.

This analysis also reveals that the shearing stress varies linearly from zero at the edge of the flange to a maximum value at the juncture with the web.

2. Compute the shear flow

Determine the shear flow at *E* and *F* (Fig. 47) by setting *Q* in $q = VQ/I$ equal to the static moment of the overhanging portion of the flange. (For convenience, use the dimensions to the centerline of the web and flange.) Thus $I = \frac{1}{12}(0.10)(16)^3 + 2(8)(0.10)(8)^2 = 137$ in^4 (5702.3 cm^4); $Q_{BE} = 5(0.10)(8) = 4.0$ in^3 (65.56 cm^3); $Q_{FG} = 3(0.10)(8) = 2.4$ in^3 (39.34 cm^3); $q_E = VQ_{BE}/I = 10,000(4.0)/137 = 292$ lb/lin in (51,137.0 N/m); $q_F = 10,000(2.4)/137 = 175$ lb/lin in (30,647.2 N/m).

3. Compute the shearing forces on the transverse section

Since the shearing stress varies linearly across the flange, $V_1 = \frac{1}{2}(292)(5) = 730$ lb (3247.0 N); $V_2 = \frac{1}{2}(175)(3) = 263$ lb (1169.8 N); $V_3 = P = 10,000$ lb (44,480 N).

4. Locate the shear center

Take moments of all forces acting on the left-hand portion of the beam with respect to a longitudinal axis through the shear center *O*. Thus $V_3 e + 16(V_2 - V_1) = 0$, or $10,000e + 16(263 - 730) = 0$; $e = 0.747$ in. (18.9738 mm).

5. Verify the computed values

Check the computed values of q_E and q_F by considering the bending stresses directly. Apply the equation $\Delta f = Vy/I$, where Δf = increase in bending stress per unit distance along the span at distance *y* from the neutral axis. Then $\Delta f = 10,000(8)/137 = 584$ lb/(sq.in.·in) (158.52 MPa/m).

In Fig. 47c, set $AB = 1$ in. (25.4 mm). Then $q_E = 584(5)(0.10) = 292$ lb/lin in (51,137.0 N/m); $q_F = 584(3)(0.10) = 175$ lb/lin in (30,647.1 N/m).

Although a particular type of beam (cantilever) was selected here for illustrative purposes and a numeric value was assigned to the vertical shear, note that the value of *e* is independent of the type of beam, form of loading, or magnitude of the vertical shear. The location of the shear center is a geometric characteristic of the transverse section.

BENDING OF A CIRCULAR FLAT PLATE

A circular steel plate 2 ft (0.61 m) in diameter and ½ in. (12.7 mm) thick, simply supported along its periphery, carries a uniform load of 20 lb/sq.in. (137.9 kPa) distributed over the entire area. Determine the maximum bending stress and deflection of this plate, using 0.25 for Poisson's ratio.

Calculation Procedure:

1. Compute the maximum stress in the plate
If the maximum deflection of the plate is less than about one-half the thickness, the effects of diaphragm behavior may be disregarded.

Compute the maximum stress, using the relation $f = (3/8)(3 + v)w(R/t)^2$, where R = plate radius, in. (mm); t = plate thickness, in. (mm); v = Poisson's ratio. Thus, $f = (3/8)(3.25)(20)(12/0.5)^2 = 14,000$ lb/sq.in. (96,530.0 kPa).

2. Compute the maximum deflection of the plate
Use the relation $y = (1 - v)(5 + v)fR^2/[2(3 + v)Et] = 0.75(5.25)(14,000)(12)^2/[2(3.25)(30 \times 10^6)(0.5)] = 0.081$ in. (2.0574 mm). Since the deflection is less than one-half the thickness, the foregoing equations are valid in this case.

BENDING OF A RECTANGULAR FLAT PLATE

A 2 × 3 ft (61.0 × 91.4 cm) rectangular plate, simply supported along its periphery, is to carry a uniform load of 8 lb/sq.in. (55.2 kPa) distributed over the entire area. If the allowable bending stress is 15,000 lb/sq.in. (103.4 MPa), what thickness of plate is required?

Calculation Procedure:

1. Select an equation for the stress in the plate
Use the approximation $f = a^2b^2w/[2(a^2 + b^2)t^2]$, where a and b denote the length of the plate sides, in. (mm).

2. Compute the required plate thickness
Solve the equation in step 1 for t. Thus $t^2 = a^2b^2w/[2(a^2 + b^2)f] = 2^2(3)^2(144)(8)/[2(2^2 + 3^2)(15,000)] = 0.106$; $t = 0.33$ in. (8.382 mm).

COMBINED BENDING AND AXIAL LOAD ANALYSIS

A post having the cross section shown in Fig. 48 carries a concentrated load of 100 kips (444.8 kN) applied at R. Determine the stress induced at each corner.

Calculation Procedure:

1. Replace the eccentric load with an equivalent system
Use a concentric load of 100 kips (444.8 kN) and two couples producing the following moments with respect to the coordinate axes:

$$M_x = 100,000(2) = 200,000 \text{ lb·in. } (25,960 \text{ N·m})$$

$$M_y = 100,000(1) = 100,000 \text{ lb·in. } (12,980 \text{ N·m})$$

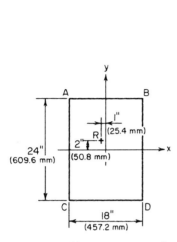

FIGURE 48. Transverse section of a post.

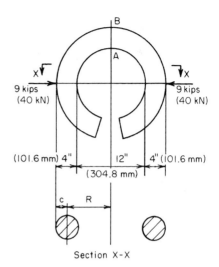

FIGURE 49. Curved member in bending.

2. Compute the section modulus
Determine the section modulus of the rectangular cross section with respect to each axis. Thus $S_x = (1/6)bd^2 = (1/6)(18)(24)^2 = 1728$ in^3 (28,321.9 cm^3); $S_y = (1/6)(24)(18)^2 = 1296$ in^3 (21,241 cm^3).

3. Compute the stresses produced
Compute the uniform stress caused by the concentric load and the stresses at the edges caused by the bending moments. Thus $f_1 = P/A = 100{,}000/[18(24)] = 231$ lb/sq.in. (1592.7 kPa); $f_x = M_x/S_x = 200{,}000/1728 = 116$ lb/sq.in. (799.8 kPa); $f_y = M_y/S_y = 100{,}000/1296 = 77$ lb/sq.in. (530.9 kPa).

4. Determine the stress at each corner
Combine the results obtained in step 3 to obtain the stress at each corner. Thus $f_A = 231 + 116 + 77 = 424$ lb/sq.in. (2923.4 kPa); $f_B = 231 + 116 - 77 = 270$ lb/sq.in. (1861.5 kPa); $f_C = 231 - 116 + 77 = 192$ lb/sq.in. (1323.8 kPa); $f_D = 231 - 116 - 77 = 38$ lb/sq.in. (262.0 kPa). These stresses are all compressive because a positive stress is considered compressive, whereas a tensile stress is negative.

5. Check the computed corner stresses
Use the following equation that applies to the special case of a rectangular cross section: $f = (P/A)(1 \pm 6e_x/d_x + 6e_y/d_y)$, where e_x and e_y = eccentricity of load with respect to the x and y axes, respectively; d_x and d_y = side of rectangle, in. (mm), normal to x and y axes, respectively. Solving for the quantities within the brackets gives $6e_x/d_x = 6(2)/24 = 0.5$; $6e_y/d_y = 6(1)/18 = 0.33$. Then $f_A = 231(1 + 0.5 + 0.33) = 424$ lb/sq.in. (2923.4 kPa); $f_B = 231(1 + 0.5 - 0.33) = 270$ lb/sq.in. (1861.5 kPa); $f_C = 231(1 - 0.5 + 0.33) = 192$ lb/sq.in. (1323.8 kPa); $f_D = 231(1 - 0.5 - 0.33) = 38$ lb/sq.in. (262.0 kPa). These results verify those computed in step 4.

FLEXURAL STRESS IN A CURVED MEMBER

The ring in Fig. 49 has an internal diameter of 12 in. (304.8 mm) and a circular cross section of 4-in (101.6-mm) diameter. Determine the normal stress at A and at B (Fig. 49).

Calculation Procedure:

1. Determine the geometrical properties of the cross section
The area of the cross section is $A = 0.7854(4)^2 = 12.56$ sq.in. (81.037 cm²); the section modulus is $S = 0.7854(2)^3 = 6.28$ in³ (102.92 cm³). With $c = 2$ in. (50.8 mm), the radius of curvature to the centroidal axis of this section is $R = 6 + 2 = 8$ in. (203.2 mm).

2. Compute the R/c ratio and determine the correction factors
Refer to a table of correction factors for curved flexural members, such as Roark—*Formulas for Stress and Strain*, and extract the correction factors at the inner and outer surface associated with the R/c ratio. Thus $R/c = 8/2 = 4$; $k_i = 1.23$; $k_o = 0.84$.

3. Determine the normal stress
Find the normal stress at A and B caused by an equivalent axial load and moment. Thus $f_A = P/A + k_i(M/S) = 9000/12.56 + 1.23(9000 \times 8)/6.28 = 14,820$ lb/sq.in. (102,183.9 kPa) compression; $f_B = 9000/12.56 - 0.84(9000 \times 8)/6.28 = 8930$ lb/sq.in. (61,572.3 kPa) tension.

SOIL PRESSURE UNDER DAM

A concrete gravity dam has the profile shown in Fig. 50. Determine the soil pressure at the toe and heel of the dam when the water surface is level with the top.

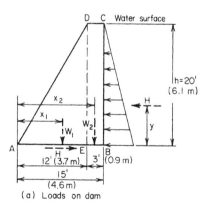

Calculation Procedure:

1. Resolve the dam into suitable elements
The soil prism underlying the dam may be regarded as a structural member subjected to simultaneous axial load and bending, the cross section of the member being identical with the bearing surface of the dam. Select a 1-ft (0.3-m) length of dam as representing the entire structure. The weight of the concrete is 150 lb/ft³ (23.56 kN/m³).

Resolve the dam into the elements *AED* and *EBCD*. Compute the weight of each element, and locate the resultant of the weight with respect to the toe. Thus $W_1 = \frac{1}{2}(12)(20)(150) = 18,000$ lb (80.06 kN); $W_2 = 3(20)(150) = 9000$ lb (40.03 kN);

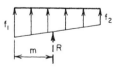

(b) Soil pressure under dam

FIGURE 50

$\Sigma W = 18{,}000 + 9000 = 27{,}000$ lb (120.10 kN). Then $x_1 = (\frac{2}{3})(12) = 8.0$ ft (2.44 m); $x^2 = 12 + 1.5 = 13.5$ ft (4.11 m).

2. Find the magnitude and location of the resultant of the hydrostatic pressure

Calling the resultant $H = \frac{1}{2}wh^2 = \frac{1}{2}(62.4)(20)^2 = 12{,}480$ lb (55.51 kN), where w = weight of water, lb/ft³ (N/m³), and h = water height, ft (m), then $y = (\frac{1}{3})(20) = 6.67$ ft (2.03 m).

3. Compute the moment of the loads with respect to the base centerline

Thus, $M = 18{,}000(8 - 7.5) + 9000(13.5 - 7.5) - 12{,}480(6.67) = 20{,}200$ lb·ft (27,391 N·m) counterclockwise.

4. Compute the section modulus of the base

Use the relation $S = (\frac{1}{6})bd^2 = (\frac{1}{6})(1)(15)^2 = 37.5$ ft³ (1.06 m³).

5. Determine the soil pressure at the dam toe and heel

Compute the soil pressure caused by the combined axial load and bending. Thus $f_1 = \Sigma W/A + M/S = 27{,}000/15 + 20{,}200/37.5 = 2339$ lb/sq.ft. (111.99 kPa); $f_2 = 1800 - 539 = 1261$ lb/sq.ft. (60.37 kPa).

6. Verify the computed results

Locate the resultant R of the trapezoidal pressure prism, and take its moment with respect to the centerline of the base. Thus $R = 27{,}000$ lb (120.10 kN); $m = (15/3)[(2 \times 1261 + 2339)/(1261 + 2339)] = 6.75$ ft (2.05 m); $M_R = 27{,}000(7.50 - 6.75) = 20{,}200$ lb·ft (27,391 N·m). Since the applied and resisting moments are numerically equal, the computed results are correct.

LOAD DISTRIBUTION IN PILE GROUP

A continuous wall is founded on three rows of piles spaced 3 ft (0.91 m) apart. The longitudinal pile spacing is 4 ft (1.21 m) in the front and center rows and 6 ft (1.82 m) in the rear row. The resultant of vertical loads on the wall is 20,000 lb/lin ft (291.87 kN/m) and lies 3 ft 3 in. (99.06 cm) from the front row. Determine the pile load in each row.

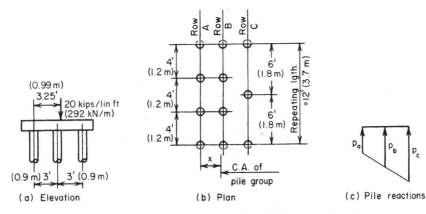

FIGURE 51

Calculation Procedure:

1. Identify the "repeating group" of piles
The concrete footing (Fig. 51a) binds the piles, causing the surface along the top of the piles to remain a plane as bending occurs. Therefore, the pile group may be regarded as a structural member subjected to axial load and bending, the cross section of the member being the aggregate of the cross sections of the piles.

Indicate the "repeating group" as shown in Fig. 51b.

2. Determine the area of the pile group and the moment of inertia
Calculate the area of the pile group, locate its centroidal axis, and find the moment of inertia. Since all the piles have the same area, set the area of a single pile equal to unity. Then $A = 3 + 3 + 2 = 8$.

Take moments with respect to row A. Thus $8x = 3(0) + 3(3) + 2(6)$; $x = 2.625$ ft (66.675 mm). Then $I = 3(2.625)^2 + 3(0.375)^2 + 2(3.375)^2 = 43.9$.

3. Compute the axial load and bending moment on the pile group
The axial load $P = 20,000(12) = 240,000$ lb (1067.5 kN); then $M = 240,000(3.25 - 2.625) = 150,000$ lb·ft (203.4 kN·m).

4. Determine the pile load in each row
Find the pile load in each row resulting from the combined axial load and moment. Thus, $P/A = 240,000/8 = 30,000$ lb (133.4 kN) per pile; then $M/I = 150,000/43.9 = 3420$. Also, $p_a = 30,000 - 3420(2.625) = 21,020$ lb (93.50 kN) per pile; $p_b = 30,000 + 3420(0.375) = 31,280$ lb (139.13 kN) per pile; $p_c = 30,000 + 3420(3.375) = 41,540$ lb (184.76 kN) per pile.

5. Verify the above results
Compute the total pile reaction, the moment of the applied load, and the pile reaction with respect to row A. Thus, $R = 3(21,020) + 3(31,280) + 2(41,540) = 239,980$ lb (1067.43 kN); then $M_a = 240,000(3.25) = 780,000$ lb·ft (1057.68 kN·m), and $M_r = 3(31,280)(3) + 2(41,540)(6) = 780,000$ lb·ft (1057.68 kN·m). Since $M_a = M_r$, the computed results are verified.

Deflection of Beams

In this handbook the slope of the elastic curve at a given section of a beam is denoted by θ, and the deflection, in inches, by y. The slope is considered positive if the section rotates in a clockwise direction under the bending loads. A downward deflection is considered positive. In all instances, the beam is understood to be prismatic, if nothing is stated to the contrary.

DOUBLE-INTEGRATION METHOD OF DETERMINING BEAM DEFLECTION

The simply supported beam in Fig. 52 is subjected to a counterclockwise moment N applied at the right-hand support. Determine the slope of the elastic curve at each support and the maximum deflection of the beam.

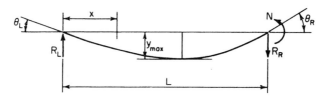

FIGURE 52. Deflection of simple beam under end moment.

Calculation Procedure:

1. Evaluate the bending moment at a given section
Make this evaluation in terms of the distance x from the left-hand support to this section. Thus $R_L = N/L$; $M = Nx/L$.

2. Write the differential equation of the elastic curve; integrate twice
Thus $EI\, d^2y/dx^2 = -M = -Nx/L$; $EI\, dy/dx = EI\theta = -Nx^2/(2L) + c_1$; $EIy = -Nx^3/(6L) + c_1 x + c_2$.

3. Evaluate the constants of integration
Apply the following boundary conditions: When $x = 0$, $y = 0$; $\therefore c_2 = 0$; when $x = L$, $y = 0$; $\therefore c_1 = NL/6$.

4. Write the slope and deflection equations
Substitute the constant values found in step 3 in the equations developed in step 2. Thus $\theta = [N/(6EIL)](L^2 - 3x^2)$; $y = [Nx/(6EIL)](L^2 - x^2)$.

5. Find the slope at the supports
Substitute the values $x = 0$, $x = L$ in the slope equation to determine the slope at the supports. Thus $\theta_L = NL/(6EI)$; $\theta_R = -NL/(3EI)$.

6. Solve for the section of maximum deflection
Set $\theta = 0$ and solve for x to locate the section of maximum deflection. Thus $L^2 - 3x^2 = 0$; $x = L/3^{0.5}$. Substituting in the deflection equation gives $y_{max} = NL^2/(9EI3^{0.5})$.

MOMENT-AREA METHOD OF DETERMINING BEAM DEFLECTION

Use the moment-area method to determine the slope of the elastic curve at each support and the maximum deflection of the beam shown in Fig. 53.

Calculation Procedure:

1. Sketch the elastic curve of the member and draw the M/(EI) diagram
Let A and B denote two points on the elastic curve of a beam. The moment-area method is based on the following theorems:

The difference between the slope at A and that at B is numerically equal to the area of the $M/(EI)$ diagram within the interval AB.

The deviation of A from a tangent to the elastic curve through B is numerically equal to the static moment of the area of the $M/(EI)$ diagram within the interval AB with respect to A. This tangential deviation is measured normal to the unstrained position of the beam.

Draw the elastic curve and the $M/(EI)$ diagram as shown in Fig. 53.

2. Calculate the deviation t_1 of B from the tangent through A

Thus, t_1 = moment of $\triangle ABC$ about $BC = [NL/(2EI)](L/3) = NL^2/(6EI)$. Also, $\theta_L = t_1/L = NL/(6EI)$.

3. Determine the right-hand slope in an analogous manner

4. Compute the distance to the section where the slope is zero

Area $\triangle AED$ = area $\triangle ABC(x/L)^2 = Nx^2/(2EIL)$; $\theta_E = \theta_L$ − area $\triangle AED = NL/(6EI) - Nx^2/(2EIL) = 0$; $x = L/3^{0.5}$.

5. Evaluate the maximum deflection

Evaluate y_{max} by calculating the deviation t_2 of A from the tangent through E' (Fig. 53). Thus area $\triangle AED = \theta_L = NL/(6EI)$; $y_{max} = t_2 = NL/(6EI)(2x/3) = [NL/(6EI)][(2L/(3 \times 3^{0.5})] = NL^2/(9EI3^{0.5})$, as before.

CONJUGATE-BEAM METHOD OF DETERMINING BEAM DEFLECTION

The overhanging beam in Fig. 54 is loaded in the manner shown. Compute the deflection at C.

Calculation Procedure:

1. Assign supports to the conjugate beam

If a conjugate beam of identical span as the given beam is loaded with the $M/(EI)$ diagram of the latter, the shear V' and bending moment M' of the conjugate beam are equal,

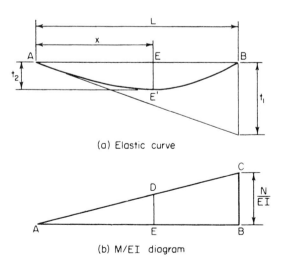

FIGURE 53

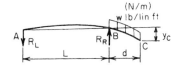

(a) Force diagram of given beam

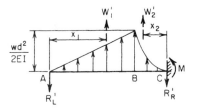

(b) Force diagram of conjugate beam

FIGURE 54. Deflection of overhanging beam.

respectively, to the slope θ and deflection y at the corresponding section of the given beam.

Assign supports to the conjugate beam that are compatible with the end conditions of the given beam. At A, the given beam has a specific slope but zero deflection. Correspondingly, the conjugate beam has a specific shear but zero moment; i.e., it is simply supported at A.

At C, the given beam has a specific slope and a specific deflection. Correspondingly, the conjugate beam has both a shear and a bending moment; i.e., it has a fixed support at C.

2. Construct the M/(EI) diagram of the given beam

Load the conjugate beam with this area. The moment at B is $-wd^2/2$; the moment varies linearly from A to B and parabolically from C to B.

3. Compute the resultant of the load in selected intervals

Compute the resultant W'_1 of the load in interval AB and the resultant W'_2 of the load in the interval BC. Locate these resultants. (Refer to the AISC *Manual* for properties of the complement of a half parabola.) Then $W'_1 = (L/2)[wd^2/(2EI)] = wd^2L/(4EI)$; $x_1 = \frac{2}{3}L$; $W'_2 = (d/3)[wd^2/(2EI)] = wd^3/(6EI)$; $x_2 = \frac{3}{4}d$.

4. Evaluate the conjugate-beam reaction

Since the given beam has zero deflection at B, the conjugate beam has zero moment at this section. Evaluate the reaction R'_L accordingly. Thus $M'_B = -R'_LL + W'_1L/3 = 0$; $R'_L W'_1/3 = wd^2L/(12EI)$.

5. Determine the deflection

Determine the deflection at C by computing M'_c. Thus $y_c = M'_c = -R'_L(L+d) + W'_1(d+L/3) + W'_2(3d/4) = wd^3(4L+3d)/(24EI)$.

UNIT-LOAD METHOD OF COMPUTING BEAM DEFLECTION

The cantilever beam in Fig. 55a carries a load that varies uniformly from w lb/lin ft at the free end to zero at the fixed end. Determine the slope and deflection of the elastic curve at the free end.

Calculation Procedure:

1. Apply a unit moment to the beam

Apply a counterclockwise unit moment at A (Fig. 55b). (This direction is selected because it is known that the end section rotates in this manner.) Let x = distance from A to given

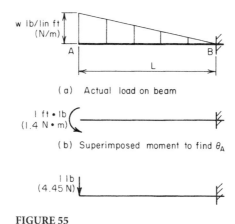

FIGURE 55

section; w_x = load intensity at the given section; M and m = bending moment at the given section induced by the actual load and by the unit moment, respectively.

2. Evaluate the moments in step 1
Evaluate M and m. By proportion, $w_x = w(L - x)/L$; $M = -(x^2/6)(2w + w_x) = -(wx^2/6)[2 + (L - x)/L] = -wx^2(3L - x)/(6L)$; $m = -1$.

3. Apply a suitable slope equation
Use the equation $\theta_A = \int_0^L [Mm/(EI)]\,dx$. Then $EI\theta_A = \int_0^L [wx^2 3L - x)/(6L)]\,dx = [w/(6L)] \times \int_0^L (3Lx^2 - x^3)\,dx = [w/(6L)](3Lx^3/3 - x^4/4)]_0^L = [w/(6L)](L^4 - L^4/4)$; thus, $\theta_A = \frac{1}{8}wL^3/(EI)$ counterclockwise. This is the slope at A.

4. Apply a unit load to the beam
Apply a unit downward load at A as shown in Fig. 55c. Let m' denote the bending moment at a given section induced by the unit load.

5. Evaluate the bending moment induced by the unit load; find the deflection
Apply $y_A = \int_0^L [Mm'/(EI)]\,dx$. Then $m' = -x$; $EIy_A = \int_0^L [wx^3(3L - x)/(6L)]\,dx = [w/(6L)] \times \int_0^L x^3(3L - x)\,dx$; $y_A = (11/120)wL^4/(EI)$.

The first equation in step 3 is a statement of the work performed by the unit moment at A as the beam deflects under the applied load. The left-hand side of this equation expresses the external work, and the right-hand side expresses the internal work. These work equations constitute a simple proof of Maxwell's theorem of reciprocal deflections, which is presented in a later calculation procedure.

DEFLECTION OF A CANTILEVER FRAME

The prismatic rigid frame $ABCD$ (Fig. 56a) carries a vertical load P at the free end. Determine the horizontal displacement of A by means of both the unit-load method and the moment-area method.

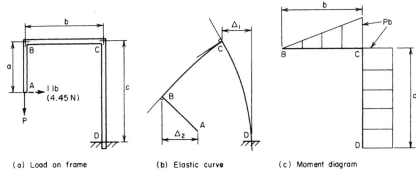

(a) Load on frame   (b) Elastic curve   (c) Moment diagram

FIGURE 56

Calculation Procedure:

1. Apply a unit horizontal load
Apply the unit horizontal load at A, directed to the right.

2. Evaluate the bending moments in each member
Let M and m denote the bending moment at a given section caused by the load P and by the unit load, respectively. Evaluate these moments in each member, considering a moment positive if it induces tension in the outer fibers of the frame. Thus:

Member AB: Let x denote the vertical distance from A to a given section. Then $M = 0$; $m = x$.

Member BC: Let x denote the horizontal distance from B to a given section. Then $M = Px$; $m = a$.

Member CD: Let x denote the vertical distance from C to a given section. Then $M = Pb$; $m = a - x$.

3. Evaluate the required deflection
Calling the required deflection Δ, we apply $\Delta = \int [Mm/(EI)]\, dx$; $EI\Delta = \int_0^b Paxdx + \int_0^c Pb(a-x)\, dx = Pax^2/2]_0^b + Pb(ax - x^2/2)]_0^c = Pab^2/2 + Pabc - Pbc^2/2$; $\Delta = [Pb/(2EI)](ab + 2ac - c^2)$.

If this value is positive, A is displaced in the direction of the unit load, i.e., to the right. Draw the elastic curve in hyperbolic fashion (Fig. 56b). The above three steps constitute the unit-load method of solving this problem.

4. Construct the bending-moment diagram
Draw the diagram as shown in Fig. 56c.

5. Compute the rotation and horizontal displacement by the moment-area method
Determine the rotation and horizontal displacement of C. (Consider only absolute values.) Since there is no rotation at D, $EI\theta_C = Pbc$; $EI\Delta_1 = Pbc^2/2$.

6. Compute the rotation of one point relative to another and the total rotation
Thus $EI\theta_{BC} = Pb^2/2$; $EI\theta_B = Pbc + Pb^2/2 = Pb(c + b/2)$. The horizontal displacement of B relative to C is infinitesimal.

7. Compute the horizontal displacement of one point relative to another
Thus, $EI\Delta_2 = EI\theta_B a = Pb(ac + ab/2)$.

8. Combine the computed displacements to obtain the absolute displacement
Thus $EI\Delta = EI(\Delta_2 - EI\Delta_1) = Pb(ac + ab/2 - c^2/2); \Delta = [Pb/(2EI)](2ac + ab - c^2)$.

Statically Indeterminate Structures

A structure is said to be *statically determinate* if its reactions and internal forces may be evaluated by applying solely the equations of equilibrium and *statically indeterminate* if such is not the case. The analysis of an indeterminate structure is performed by combining the equations of equilibrium with the known characteristics of the deformation of the structure.

SHEAR AND BENDING MOMENT OF A BEAM ON A YIELDING SUPPORT

The beam in Fig. 57a has an EI value of 35×10^9 lb·sq.in. (100,429 kN·m²) and bears on a spring at B that has a constant of 100 kips/in (175,126.8 kN/m); i.e., a force of 100 kips (444.8 kN) will compress the spring 1 in. (25.4 mm). Neglecting the weight of the member, construct the shear and bending-moment diagrams.

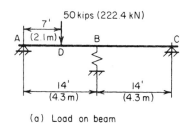

(a) Load on beam

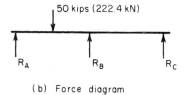

(b) Force diagram

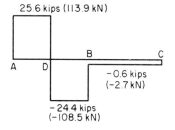

(c) Shear diagram

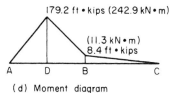

(d) Moment diagram

FIGURE 57

Calculation Procedure:

1. Draw the free-body diagram of the beam
Draw the diagram in Fig. 57b. Consider this as a simply supported member carrying a 50-kip (222.4-kN) load at D and an upward load R_B at its center.

2. Evaluate the deflection
Evaluate the deflection at B by applying the equations presented for cases 7 and 8 in the AISC *Manual*. With respect to the 50-kip (222.4-kN) load, $b = 7$ ft (2.1 m) and $x = 14$ ft (4.3 m). If y is in inches and R_B is in pounds, $y = 50,000(7)(14)(28^2 - 7^2 - 14^2)1728/[6(35)(10)^9 28] - R_B(28)^3 1728/[48(35)(10)^9] = 0.776 - (2.26/10^5)R_B$.

3. Express the deflection in terms of the spring constant
The deflection at B is, by proportion, $y/1 = R_B/100,000$; $y = R_B/100,000$.

4. Equate the two deflection expressions, and solve for the upward load
Thus $R_B/10^5 = 0.776 - (2.26/10^5)R_B$; $R_B = 0.776(10)^5/3.26 = 23,800$ lb (105,862.4 N).

5. Calculate the reactions R_A and R_C by taking moments
We have $\Sigma M_C = 28R_A - 50,000(21) + 23,800(14) = 0$; $R_A = 25,600$ lb (113,868.8 N); $\Sigma M_A = 50,000(7) - 23,800(14) - 28R_C = 0$; $R_C = 600$ lb (2668.8 N).

6. Construct the shear and moment diagrams
Construct these diagrams as shown in Fig. 57. Then $M_D = 7(25,600) = 179,200$ lb·ft (242,960 N·m); $M_B = 179,200 - 7(24,400) = 8400$ lb·ft (11,390.4 N·m).

MAXIMUM BENDING STRESS IN BEAMS JOINTLY SUPPORTING A LOAD

In Fig. 58a, a W16 × 40 beam and a W12 × 31 beam cross each other at the vertical line V, the bottom of the 16-in. (406.4-mm) beam being ⅜ in. (9.53 mm) above the top of the 12-in. (304.8-mm) beam before the load is applied. Both members are simply

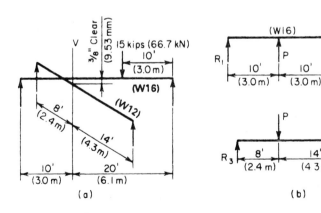

FIGURE 58. Load carried by two beams.

supported. A column bearing on the 16-in. (406.4-mm) beam transmits a load of 15 kips (66.72 kN) at the indicated location. Compute the maximum bending stress in the 12-in. (304.8-mm) beam.

Calculation Procedure:

1. Determine whether the upper beam engages the lower beam
To ascertain whether the upper beam engages the lower one as it deflects under the 15-kip (66.72-kN) load, compute the deflection of the 16-in. (406.4-mm) beam at V if the 12-in. (304.8-mm) beam were absent. This distance is 0.74 in. (18.80 mm). Consequently, the gap between the members is closed, and the two beams share the load.

2. Draw a free-body diagram of each member
Let P denote the load transmitted to the 12-in. (304.8-mm) beam by the 16-in. (406.4-mm) beam [or the reaction of the 12-in. (304.8-mm) beam on the 16-in. (406.4-mm) beam]. Draw, in Fig. 58b, a free-body diagram of each member.

3. Evaluate the deflection of the beams
Evaluate, in terms of P, the deflections y_{12} and y_{16} of the 12-in. (304.8-mm) and 16-in. (406.4-mm) beams, respectively, at line V.

4. Express the relationship between the two deflections
Thus, $y_{12} = y_{16} - 0.375$.

5. Replace the deflections in step 4 with their values as obtained in step 3
After substituting these deflections, solve for P.

6. Compute the reactions of the lower beam
Once the reactions of the lower beam are computed, obtain the maximum bending moment. Then compute the corresponding flexural stress.

THEOREM OF THREE MOMENTS

For the two-span beam in Figs. 59 and 60, compute the reactions at the supports. Apply the theorem of three moments to arrive at the results.

Calculation Procedure:

1. Using the bending-moment equation, determine M_B
Figure 59 represents a general case. For a prismatic beam, the bending moments at the three successive supports are related by $M_1 L_1 + 2M_2(L_1 + L_2) + M_3 L_2 - \frac{1}{4} w_1 L_1^3 - \frac{1}{4} w_2 L_2^3 - P_1 L_1^2(k_1 - k_1^3) - P_2 L_2^2(k_2 - k_2^3)$. Substituting in this equation gives $M_1 = M_3 = 0$; $L_1 = 10$ ft (3.0 m); $L_2 = 15$ ft (4.6 m); $w_1 = 2$ kips/lin ft (29.2 kN/m); $w_2 = 3$ kips/lin ft (43.8 kN/m); $P_1 = 6$ kips (26.7 kN); $P_2 = 10$ kips (44.5 kN); $k_1 = 0.5$; $k_2 = 0.4$; $2M_B(10 + 15) = \frac{1}{4}(2)(10)^3 - \frac{1}{4}(3)(15)^3 - 6(10)^2(0.5 - 0.125) - 10(15)^2(0.4 - 0.064)$; $M_B = -80.2$ ft·kips (-108.8 kN·m).

2. Draw a free-body diagram of each span
Figure 59 shows the free-body diagrams.

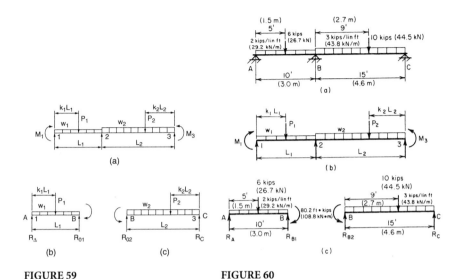

FIGURE 59   FIGURE 60

3. Take moments with respect to each support to find the reactions

Span AB: $\Sigma M_A = 6(5) + 2(10)(5) + 80.2 - 10R_{B1} = 0$; $R_{B1} = 21.02$ kips (93.496 kN); $\Sigma M_B = 10R_A - 6(5) - 2(10)(5) + 80.2 = 0$; $R_A = 4.98$ kips (22.151 kN).

Span BC: $\Sigma M_B = -80.2 + 10(9) + 3(15)(7.5) - 15R_C = 0$; $R_C = 23.15$ kips (102.971 kN); $\Sigma M_C = 15R_{B2} - 80.2 - 10(6) - 3(15)(7.5) = 0$; $R_{B2} = 31.85$ kips (144.668 kN); $R_B = 21.02 + 31.85 = 52.87$ kips (235.165 kN).

THEOREM OF THREE MOMENTS: BEAM WITH OVERHANG AND FIXED END

Determine the reactions at the supports of the continuous beam in Fig. 60a. Use the theorem of three moments.

Calculation Procedure:

1. Transform the given beam to one amenable to analysis by the theorem of three moments

Perform the following operations to transform the beam:

a. Remove the span AB, and introduce the shear V_B and moment M_B that the load on AB induces at B, as shown in Fig. 60b.

b. Remove the fixed support at D and add the span DE of zero length, with a hinged support at E.

For the interval BD, the transformed beam is then identical in every respect with the actual beam.

2. Apply the equation for the theorem of three moments

Consider span BC as span 1 and CD as span 2. For the 5-kip (22.2-kN) load, $k_2 = 12/16 = 0.75$; for the 10-kip (44.5-kN) load, $k_2 = 8/16 = 0.5$. Then $-12(10) + 2M_C(10 + 16) + 16M_D = \frac{1}{4}(4)(10)^3 - 5(16)^2(0.75 - 0.422) - 10(16)^2(0.5 - 0.125)$. Simplifying gives $13M_C + 4M_D = -565.0$, Eq. a.

3. Apply the moment equation again

Considering CD as span 1 and DE as span 2, apply the moment equation again. Or, for the 5-kip (22.2-kN) load, $k_1 = 0.25$; for the 10-kip (44.5-kN) load, $k_1 = 0.5$. Then $16M_C + 2M_D(16 + 0) = -5(16)^2(0.25 - 0.016) - 10(16)^2(0.50 - 0.125)$. Simplifying yields $M_C + 2M_D = -78.7$, Eq. b.

4. Solve the moment equations

Solving Eqs. a and b gives $M_C = -37.1$ ft·kips (-50.30 kN·m); $M_D = -20.8$ ft·kips (-28.20 kN·m).

5. Determine the reactions by using a free-body diagram

Find the reactions by drawing a free-body diagram of each span and taking moments with respect to each support. Thus $R_B = 20.5$ kips (91.18 kN); $R_C = 32.3$ kips (143.67 kN); $R_D = 5.2$ kips (23.12 kN).

BENDING-MOMENT DETERMINATION BY MOMENT DISTRIBUTION

Using moment distribution, determine the bending moments at the supports of the member in Fig. 61. The beams are rigidly joined at the supports and are composed of the same material.

Calculation Procedure:

1. Calculate the flexural stiffness of each span

Using K to denote the flexural stiffness, we see that $K = I/L$ if the far end remains fixed during moment distribution; $K = 0.75I/L$ if the far end remains hinged during moment distribution. Then $K_{AB} = 270/18 = 15$; $K_{BC} = 192/12 = 16$; $K_{CD} = 0.75(240/20) = 9$. Record all the values on the drawing as they are obtained.

2. For each span, calculate the required fixed-end moments at those supports that will be considered fixed

These are the external moments with respect to the span; a clockwise moment is considered positive. (For additional data, refer to cases 14 and 15 in the AISC Manual.) Then $M_{AB} = -wL^2/12 = -2(18)^2/12 = -54.0$ ft·kips (-73.2 kN·m); $M_{BA} = +54.0$ ft·kips (73.22 kN·m). Similarly, $M_{BC} = -48.0$ ft·kips (-65.1 kN·m); $M_{CB} = +48.0$ ft·kips (65.1 kN·m); $M_{CD} = -24(15)(5)(15 + 20)/[2(20)^2] = -78.8$ ft·kips (-106.85 kN·m).

3. Calculate the unbalanced moments

Computing the unbalanced moments at B and C yields the following: At B, $+54.0 - 48.0 = +6.0$ ft·kips (8.14 kN·m); at C, $+48.0 - 78.8 = -30.8$ ft·kips (-41.76 kN·m).

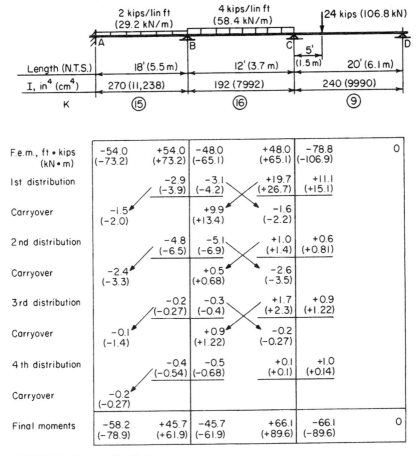

FIGURE 61. Moment distribution.

4. Apply balancing moments; distribute them in proportion to the stiffness of the adjoining spans

Apply the balancing moments at B and C, and distribute them to the two adjoining spans in proportion to their stiffness. Thus $M_{BA} = -6.0(15/31) = -2.9$ ft·kips (-3.93 kN·m); $M_{BC}=-6.0(16/31)=-3.1$ ft·kips (-4.20 kN·m); $M_{CB}=+30.8(16/25)=+19.7$ ft·kips (26.71 kN m); $M_{CD}=+30.8(9/25)=+11.1$ ft·kips (15.05 kN·m).

5. Perform the "carry-over" operation for each span

To do this, take one-half the distributed moment applied at one end of the span, and add this to the moment at the far end if that end is considered to be fixed during moment distribution.

6. Perform the second cycle of moment balancing and distribution
Thus $M_{BA} = -9.9(15/31) = -4.8$; $M_{BC} = -9.9(16/31) = -5.1$; $M_{CB} = +1.6(16/25) = +1.0$; $M_{CD} = +1.6(9/25) = +0.6$.

7. Continue the foregoing procedure until the carry-over moments become negligible
Total the results to obtain the following bending moments: $M_A = -58.2$ ft·kips (-78.91 kN·m); $M_B = -45.7$ ft·kips (-61.96 kN·m); $M_C = -66.1$ ft·kips (-89.63 kN·m).

ANALYSIS OF A STATICALLY INDETERMINATE TRUSS

Determine the internal forces of the truss in Fig. 62a. The cross-sectional areas of the members are given in Table 6.

Calculation Procedure:

1. Test the structure for static determinateness
Apply the following criterion. Let j = number of joints; m = number of members; r = number of reactions. Then if $2j = m + r$, the truss is statically determinate; if $2j < m + r$, the truss is statically indeterminate and the deficiency represents the degree of indeterminateness.

In this truss, $j = 6$, $m = 10$, $r = 3$, consisting of a vertical reaction at A and D and a horizontal reaction at D. Thus $2j = 12$; $m + r = 13$. The truss is therefore statically indeterminate to the first degree; i.e., there is *one* redundant member.

The method of analysis comprises the following steps: Assume a value for the internal force in a particular member, and calculate the relative displacement Δ_i, of the two ends of that member caused solely by this force. Now remove this member to secure a determinate truss, and calculate the relative displacement Δ_a caused solely by the applied loads. The true internal force is of such magnitude that $\Delta_i = -\Delta_a$.

2. Assume a unit force for one member
Assume for convenience that the force in BF is 1-kip (4.45-kN) tension. Remove this member, and replace it with the assumed 1-kip (4.45-kN) force that it exerts at joints B and F, as shown in Fig. 62b.

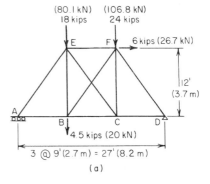

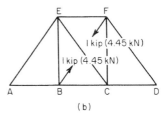

FIGURE 62. Statically indeterminate truss.

TABLE 6. Forces in Truss Members (Fig. 62)

Member	A, sq.in. (cm²)	L, in. (mm)	U, kips (kN)	S, kips (kN)	U²L/A	SUL/A	S', kips (kN)
AB	5 (32.2)	108 (2,743.2)	0 (0)	+15.25 (+67.832)	0 (0)	0 (0)	+15.25 (+67.832)
BC	5 (32.2)	108 (2,743.2)	−0.60 (−2.668)	+15.25 (+67.832)	+7.8 (+615.54)	−197.6 (−15,417.78)	+14.07 (+62.588)
CD	5 (32.2)	108 (2,743.2)	0 (0)	+13.63 (+60.626)	0 (0)	0 (0)	+13.63 (+60.626)
EF	4 (25.8)	108 (2,743.2)	−0.60 (−2.688)	−13.63 (−60.626)	+9.7 (+756.84)	+220.8 (+17,198.18)	−14.81 (−65.874)
BE	4 (25.8)	144 (3,657.6)	−0.80 (−3.558)	+4.50 (+20.016)	+23.0 (+1,794.68)	−129.6 (−10,096.24)	+2.92 (+12.988)
CF	4 (25.8)	144 (3,657.6)	−0.80 (−3.558)	+2.17 (+9.952)	+23.0 (+1,794.68)	−62.5 (−4,868.55)	+0.59 (+2.624)
AE	6 (38.7)	180 (4,572.0)	0 (0)	−25.42 (−113.068)	0 (0)	0 (0)	−25.42 (−113.068)
BF	5 (32.2)	180 (4,572.0)	+1.00 (+4.448)	0 (0)	+36.0 (+2,809.18)	0 (0)	+1.97 (+8.762)
CE	5 (32.2)	180 (4,572.0)	+1.00 (+4.448)	−2.71 (−9.652)	+36.0 (+2,809.18)	−97.6 (−6,095.82)	−0.74 (−3.291)
DF	6 (38.7)	180 (4,572.0)	0 (0)	−32.71 (−145.494)	0 (0)	0 (0)	−32.71 (−145.494)
Total					+135.5 (+10,580.1)	−266.5 (−19,280.2)	

3. Calculate the force induced in each member solely by the unit force

Calling the induced force U, produced solely by the unit tension in BF, record the results in Table 6, considering tensile forces as positive and compressive forces as negative.

4. Calculate the force induced in each member solely by the applied loads

With BF eliminated, calculate the force S induced in each member solely by the applied loads.

5. Evaluate the true force in the selected member

Use the relation $BF = -[\Sigma SUL/(AE)]/[\Sigma U^2L/(AE)]$. The numerator represents Δ_a; the denominator represents Δ_i for a 1-kip (4.45-kN) tensile force in BF. Since E is constant, it cancels. Substituting the values in Table 6 gives $BF = -(-266.5/135.5) = 1.97$ kips (8.76 kN). The positive result confirms the assumption that BF is tensile.

6. Evaluate the true force in each member

Use the relation $S' = S + 1.97 U$, where S' = true force. The results are shown in Table 6.

Moving Loads and Influence Lines

ANALYSIS OF BEAM CARRYING MOVING CONCENTRATED LOADS

The loads shown in Fig. 63a traverse a beam of 40-ft (12.2-m) simple span while their spacing remains constant. Determine the maximum bending moment and maximum shear induced in the beam during transit of these loads. Disregard the weight of the beam.

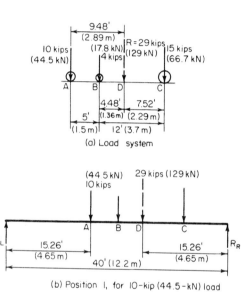

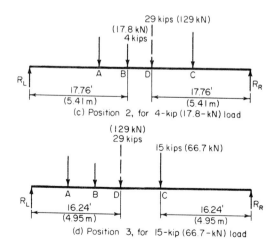

FIGURE 63

Calculation Procedure:

1. Determine the magnitude of the resultant and its location
Since the member carries only concentrated loads, the maximum moment at any instant occurs under one of these loads. Thus, the problem is to determine the position of the load system that causes the *absolute* maximum moment.

The magnitude of the resultant R is $R = 10 + 4 + 15 = 29$ kips (129.0 kN). To determine the location of R, take moments with respect to A (Fig. 63). Thus $\Sigma M_A = 29AD = 4(5) + 15(17)$, or $AD = 9.48$ ft (2.890 m).

2. Assume several trial load positions
Assume that the maximum moment occurs under the 10-kip (44.5-kN) load. Place the system in the position shown in Fig. 63b, with the 10-kip (44.5-kN) load as far from the adjacent support as the resultant is from the other support. Repeat this procedure for the two remaining loads.

3. Determine the support reactions for the trial load positions
For these three trial positions, calculate the reaction at the support adjacent to the load under consideration. Determine whether the vertical shear is zero or changes sign at this load. Thus, for position 1: $R_L = 29(15.26)/40 = 11.06$ kips (49.194 kN). Since the shear does not change sign at the 10-kip (44.5-kN) load, this position lacks significance.

Position 2: $R_L = 29(17.76)/40 = 12.88$ kips (57.290 kN). The shear changes sign at the 4-kip (17.8-kN) load.

Position 3: $R_R = 29(16.24)/40 = 11.77$ kips (52.352 kN). The shear changes sign at the 15-kip (66.7-kN) load.

4. Compute the maximum bending moment associated with positions having a change in the shear sign
This applies to positions 2 and 3. The absolute maximum moment is the larger of these values. Thus, for position 2: $M = 12.88(17.76) - 10(5) = 178.7$ ft·kips (242.32 kN·m). Position 3: $M = 11.77(16.24) = 191.1$ ft·kips (259.13 kN·m). Thus, $M_{max} = 191.1$ ft·kips (259.13 kN·m).

5. Determine the absolute maximum shear
For absolute maximum shear, place the 15-kip (66.7-kN) load an infinitesimal distance to the left of the right-hand support. Then $V_{max} = 29(40 - 7.52)/40 = 23.5$ kips (104.53 kN).

When the load spacing is large in relation to the beam span, the absolute maximum moment may occur when only part of the load system is on the span. This possibility requires careful investigation.

INFLUENCE LINE FOR SHEAR IN A BRIDGE TRUSS

The Pratt truss in Fig. 64a supports a bridge at its bottom chord. Draw the influence line for shear in panel cd caused by a moving load traversing the bridge floor.

Calculation Procedure:

1. Compute the shear in the panel being considered with a unit load to the right of the panel
Cut the truss at section YY. The algebraic sum of vertical forces acting on the truss at panel points to the left of YY is termed the shear in panel cd.

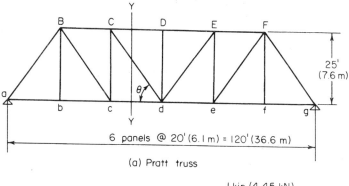

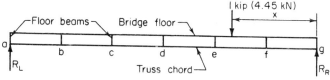

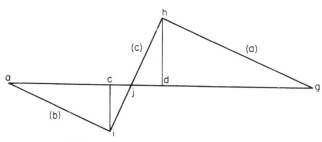

FIGURE 64

Consider that a moving load traverses the bridge floor from right to left and that the portion of the load carried by the given truss is 1 kip (4.45 kN). This unit load is transmitted to the truss as concentrated loads at two adjacent bottom-chord panel points, the latter being components of the unit load. Let x denote the instantaneous distance from the right-hand support to the moving load.

Place the unit load to the right of d, as shown in Fig. 64b, and compute the shear V_{cd} in panel cd. The truss reactions may be obtained by considering the unit load itself rather than its panel-point components. Thus: $R_L = x/120$; $V_{cd} = R_L = x/120$, Eq. a.

2. Compute the panel shear with the unit load to the left of the panel considered

Placing the unit load to the left of c yields $V_{cd} = R_L - 1 = x/120 - 1$, Eq. b.

3. Determine the panel shear with the unit load within the panel

Place the unit load within panel cd. Determine the panel-point load P_c at c, and compute V_{cd}. Thus $P_c = (x - 60)/20 = x/20 - 3$; $V_{cd} = R_L - P_c = x/120 - (x/20 - 3) = -x/24 + 3$, Eq. c.

4. Construct a diagram representing the shear associated with every position of the unit load

Apply the foregoing equations to represent the value of V_{cd} associated with every position of the unit load. This diagram, Fig. 64c, is termed an *influence line*. The point j at which this line intersects the base is referred to as the *neutral point*.

5. Compute the slope of each segment of the influence line

Line a, $dV_{cd}/dx = 1/120$; line b, $dV_{cd}/dx = 1/120$; line c, $dV_{cd}/dx = -1/24$. Lines a and b are therefore parallel because they have the same slope.

FORCE IN TRUSS DIAGONAL CAUSED BY A MOVING UNIFORM LOAD

The bridge floor in Fig. 64a carries a moving uniformly distributed load. The portion of the load transmitted to the given truss is 2.3 kips/lin ft (33.57 kN/m). Determine the limiting values of the force induced in member Cd by this load.

Calculation Procedure:

1. Locate the neutral point, and compute dh

The force in Cd is a function of V_{cd}. Locate the neutral point j in Fig. 64c and compute dh. From Eq. c of the previous calculation procedure, $V_{cd} = -jg/24 + S = 0$; $jg = 72$ ft (21.9 m). From Eq. a of the previous procedure, $dh = 60/120 = 0.5$.

2. Determine the maximum shear

To secure the maximum value of V_{cd}, apply uniform load continuously in the interval jg. Compute V_{cd} by multiplying the area under the influence line by the intensity of the applied load. Thus, $V_{cd} = \frac{1}{2}(72)(0.5)(2.3) = 41.4$ kips (184.15 kN).

3. Determine the maximum force in the member

Use the relation $Cd_{max} = V_{cd}(\csc \theta)$, where $\csc \theta = [(20^2 + 25^2)/25^2]^{0.5} = 1.28$. Then $Cd_{max} = 41.4(1.28) = 53.0$-kip (235.74-kN) tension.

4. Determine the minimum force in the member

To secure the minimum value of V_{cd}, apply uniform load continuously in the interval aj. Perform the final calculation by proportion. Thus, $Cd_{min}/Cd_{max} = \text{area } aij/\text{area } jhg = -(\frac{2}{3})^2 = 9$. Then $Cd_{min} = -(4/9)(53.0) = 23.6$-kip (104.97-kN) compression.

FORCE IN TRUSS DIAGONAL CAUSED BY MOVING CONCENTRATED LOADS

The truss in Fig. 65a supports a bridge that transmits the moving-load system shown in Fig. 65b to its bottom chord. Determine the maximum tensile force in De.

Calculation Procedure:

1. Locate the resultant of the load system

The force in De (Fig. 65) is a function of the shear in panel de. This shear is calculated without recourse to a set rule in order to show the principles involved in designing for moving loads.

STRUCTURAL STEEL DESIGN   1.107

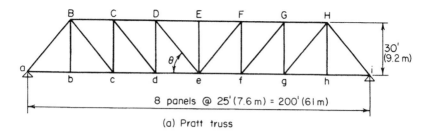

(a) Pratt truss

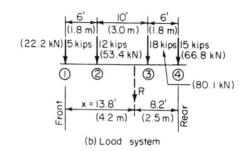

(b) Load system

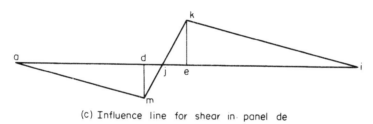

(c) Influence line for shear in panel de

FIGURE 65

To locate the resultant of the load system, take moments with respect to load 1. Thus, $R = 50$ kips (222.4 kN). Then $\Sigma M_1 = 12(6) + 18(16) + 15(22) = 50x$; $x = 13.8$ ft (4.21 m).

2. Construct the influence line for V_{de}

In Fig. 49c, draw the influence line for V_{de}. Assume right-to-left locomotion, and express the slope of each segment of the influence line. Thus slope of ik = slope of ma = 1/200; slope of $km = -7/200$.

3. Assume a load position, and determine whether V_{de} increases or decreases

Consider that load 1 lies within panel de and the remaining loads lie to the right of this panel. From the slope of the influence line, ascertain whether V_{de} increases or decreases as the system is displaced to the left. Thus $dV_{de}/dx = 5(-7/200) + 45(1/200) > 0$; $\therefore V_{de}$ increases.

4. Repeat the foregoing calculation with other assumed load positions

Consider that loads 1 and 2 lie within the panel de and the remaining loads lie to the right of this panel. Repeat the foregoing calculation. Thus $dV_{de}/dx = 17(-7/200) + 33(1/200) < 0$; $\therefore V_{de}$ decreases.

From these results it is concluded that as the system moves from right to left, V_{de} is maximum at the instant that load 2 is at e.

5. Place the system in the position thus established, and compute V_{de}

Thus, $R_L = 50(100 + 6 - 13.8)/200 = 23.1$ kips (102.75 kN). The load at panel point d is $P_d = 5(6)/25$ 1.2 kips (5.34 kN); $V_{de} = 23.1 - 1.2 = 21.9$ kips (97.41 kN).

6. Assume left-to-right locomotion; proceed as in step 3

Consider that load 4 is within panel de and the remaining loads are to the right of this panel. Proceeding as in step 3, we find $dV_{de}/dx = 15(7/200) + 35(-1/200) > 0$.

So, as the system moves from left to right, V_{de} is maximum at the instant that load 4 is at e.

7. Place the system in the position thus established, and compute V_{de}

Thus $V_{de} = R_L = [50(100 - 8.2)1/200 = 23.0$ kips (102.30 kN); $\therefore V_{de,max} = 23.0$ kips (102.30 kN).

8. Compute the maximum tensile force in De

Using the same relation as in step 3 of the previous calculation procedure, we find csc $\theta = [(25^2 + 30^2)/30^2]^{0.5} = 1.30$; then $De = 23.0(1.30) = 29.9$-kip (133.00-kN) tension.

INFLUENCE LINE FOR BENDING MOMENT IN BRIDGE TRUSS

The Warren truss in Fig. 66a supports a bridge at its top chord. Draw the influence line for the bending moment at b caused by a moving load traversing the bridge floor.

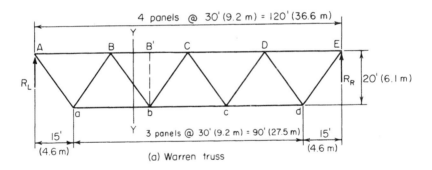

(a) Warren truss

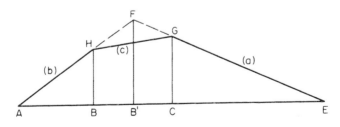

(b) Influence line for bending moment at b

FIGURE 66

Calculation Procedure:

1. Place the unit load in position, and compute the bending moment

The moment of all forces acting on the truss at panel points to the left of b with respect to b is termed the bending moment at that point. Assume that the load transmitted to the given truss is 1 kip (4.45 kN), and let x denote the instantaneous distance from the right-hand support to the moving load.

Place the unit load to the right of C, and compute the bending moment M_b. Thus $R_L = x/120$; $M_b = 45R_L = 3x/8$, Eq. a.

2. Place the unit load on the other side and compute the bending moment

Placing the unit load to the left of B and computing M_b, $M_b = 45R_L - (x - 75) = -5x/8 + 75$, Eq. b.

3. Place the unit load within the panel; compute the panel-point load and bending moment

Place the unit load within panel BC. Determine the panel-point load P_B and compute M_b. Thus $= P_B(x - 60)/30 = x/30 - 2$; $M_b = 45R_L - 15P_B = 3x/8 - 15(x/30 - 2) = -x/8 + 30$, Eq. c.

4. Applying the foregoing equations, draw the influence line

Figure 50b shows the influence line for M_b. Computing the significant values yields $CG = (\frac{3}{8})(60) = 22.50$ ft·kips (30.51 kN·m); $BH = -(5/8)(90) + 75 = 18.75$ ft·kips (25.425 kN·m).

5. Compute the slope of each segment of the influence line

This computation is made for subsequent reference. Thus, line a, $dM_b/dx = \frac{3}{8}$; line b, $dM_b/dx = -\frac{5}{8}$; line c, $dM_b/dx = -\frac{1}{8}$.

FORCE IN TRUSS CHORD CAUSED BY MOVING CONCENTRATED LOADS

The truss in Fig. 66a carries the moving-load system shown in Fig. 67. Determine the maximum force induced in member BC during transit of the loads.

Calculation Procedure:

1. Assume that locomotion proceeds from right to left, and compute the bending moment

The force in BC is a function of the bending moment M_b at b. Refer to the previous calculation procedure for the slope of each segment of the influence line. Study of these slopes shows that M_b increases as the load system moves until the rear load is at C, the front load being 14 ft (4.3 m) to the left of C. Calculate the value of M_b corresponding to this load disposition by applying the computed properties of the influence line. Thus, $M_b = 22.50(24) + (22.50 - 1/8 \times 14)(6) = 664.5$ ft·kips (901.06 kN·m).

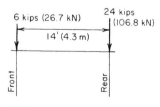

FIGURE 67

2. Assume that locomotion proceeds from left to right, and compute the bending moment

Study shows that M_b increases as the system moves until the rear load is at C, the front load being 14 ft (4.3 m) to the right of C. Calculate the corresponding value of M_b. Thus, $M_b = 22.50(24) + (22.50 - \frac{3}{8} \times 14)(6) = 6435$ ft·kips (872.59 kN·m). $\therefore M_{b,max} = 664.5$ ft·kips (901.06 kN·m).

3. Determine the maximum force in the member

Cut the truss at plane YY. Determine the maximum force in BC by considering the equilibrium of the left part of the structure. Thus, $\Sigma M_b = M_b - 20BC = 0$; $BC = 664.5/20 = 33.2$-kips (147.67-kN) compression.

INFLUENCE LINE FOR BENDING MOMENT IN THREE-HINGED ARCH

The arch in Fig. 68a is hinged at A, B, and C. Draw the influence line for bending moment at D, and locate the neutral point.

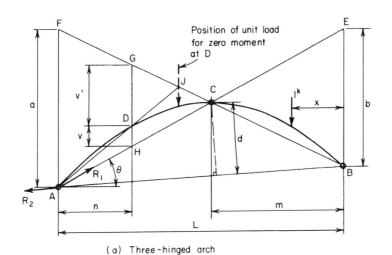

(a) Three-hinged arch

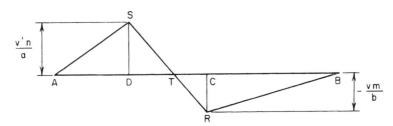

(b) Influence line for bending moment at D

FIGURE 68

Calculation Procedure:

1. Start the graphical construction
Draw a line through A and C, intersecting the vertical line through B at E. Draw a line through B and C, intersecting the vertical line through A and F. Draw the vertical line GH through D.

Let θ denote the angle between AE and the horizontal lines through B and D perpendicular to AE (omitted for clarity) make an angle θ with the vertical.

2. Resolve the reaction into components
Resolve the reaction at A into the components R_1 and R_2 acting along AE and AB, respectively (Fig. 68).

3. Determine the value of the first reaction
Let x denote the horizontal distance from the right-hand support to the unit load, where x has any value between 0 and L. Evaluate R_1 by equating the bending moment at B to zero. Thus $M_B = R_1 b \cos\theta - x = 0$; or $= R_1 = x/(b \cos\theta)$.

4. Evaluate the second reaction
Place the unit load within the interval CB. Evaluate R_2 by equating the bending moment at C to zero. Thus $M_C = R_2 d = 0$; $\therefore R_2 = 0$.

5. Calculate the bending moment at D when the unit load lies within the interval CB
Thus, $M_D = -R_1 v \cos\theta = -[(v \cos\theta)/(b \cos\theta)]x$, or $M_D = -vx/b$, Eq. a. When $x = m$, $M_D = -vm/b$.

6. Place the unit load in a new position, and determine the bending moment
Place the unit load within the interval AD. Working from the right-hand support, proceed in an analogous manner to arrive at the following result: $M_D = v'(L - x)/a$, Eq. b. When $x = L - n$, $M_D = v'n/a$.

7. Place the unit load within another interval, and evaluate the second reaction
Place the unit load within the interval DC, and evaluate R_2. Thus $M_C = R_2 d - (x - m) = 0$, or $R_2 = (x - m)/d$.

Since both R_1 and R_2 vary linearly with respect to x, it follows that M_D is also a linear function of x.

8. Complete the influence line
In Fig. 68b, draw lines BR and AS to represent Eqs. a and b, respectively. Draw the straight line SR, thus completing the influence line. The point T at which this line intersects the base is termed the neutral point.

9. Locate the neutral point
To locate T, draw a line through A and D in Fig. 68a intersecting BF at J. The neutral point in the influence line lies vertically below J; that is, M_D is zero when the action line of the unit load passes through J.

The proof is as follows: Since $M_D = 0$ and there are no applied loads in the interval AD, it follows that the total reaction at A is directed along AD. Similarly, since $M_C = 0$ and there are no applied loads in the interval CB, it follows that the total reaction at B is directed along BC. Because the unit load and the two reactions constitute a balanced system of forces, they are collinear. Therefore, J lies on the action line of the unit load.

Alternatively, the location of the neutral point may be established by applying the geometric properties of the influence line.

DEFLECTION OF A BEAM UNDER MOVING LOADS

The moving-load system in Fig. 69a traverses a beam on a simple span of 40 ft (12.2 m). What disposition of the system will cause the maximum deflection at midspan?

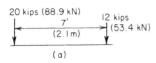

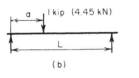

FIGURE 69

Calculation Procedure:

1. Develop the equations for the midspan deflection under a unit load

The maximum deflection will manifestly occur when the two loads lie on opposite sides of the centerline of the span. In calculating the deflection at midspan caused by a load applied at any point on the span, it is advantageous to apply Maxwell's theorem of reciprocal deflections, which states the following: *The deflection at A caused by a load at B equals the deflection at B caused by this load at A.*

In Fig. 69b, consider the beam on a simple span L to carry a unit load applied at a distance a from the left-hand support. By referring to case 7 of the AISC *Manual* and applying the principle of reciprocal deflections, derive the following equations for the midspan deflection under the unit load: When $a < L/2$, $y = (3L^2 a - 4a^3)/(48EI)$. When $a < L/2$, $y = [3L^2(L-a) - 4(L-a)^3]/(48EI)$.

2. Position the system for purposes of analysis

Position the system in such a manner that the 20-kip (89.0-kN) load lies to the left of center and the 12-kip (53.4-kN) load lies to the right of center. For the 20-kip (89.0-kN) load, set $a = x$. For the 12-kip (53.4-kN) load, $a = x + 7$; $L - a = 40 - (x + 7) = 33 - x$.

3. Express the total midspan deflection in terms of x

Substitute in the preceding equations. Combining all constants into a single term k, we find $ky = 20(3) \times 40^2 x - 4x^3) + 12[3 \times 40^2(33-x) - 4(33-x)^3]$.

4. Solve for the unknown distance

Set $dy/dx = 0$ and solve for x. Thus, $x = 17.46$ ft (5.321 m).

For maximum deflection, position the load system with the 20-kip (89.0-kN) load 17.46 ft (5.321 m) from the left-hand support.

Riveted and Welded Connections

In the design of riveted and welded connections in this handbook, the American Institute of Steel Construction *Specification for the Design, Fabrication and Erection of Structural Steel for Buildings* is applied. This is presented in Part 5 of the *Manual of Steel Construction*.

The structural members considered here are made of ASTM A36 steel having a yield-point stress of 36,000 lb/sq.in. (248,220 kPa). (The yield-point stress is denoted by F_y in the *Specification*.) All connections considered here are made with A141 hot-driven rivets or fillet welds of A233 class E60 series electrodes.

From the *Specification*, the allowable stresses are as follows: Tensile stress in connected member, 22,000 lb/sq.in. (151,690.0 kPa); shearing stress in rivet, 15,000 lb/sq.in. (103,425.0 kPa); bearing stress on projected area of rivet, 48,500 lb/sq.in. (334,408.0 kPa); stress on throat of fillet weld, 13,600 lb/sq.in. (93,772.0 kPa).

Let n denote the number of sixteenths included in the size of a fillet weld. For example, for a ⅜-in. (9.53-mm) weld, $n = 6$. Then weld size $= n/16$. And throat area per linear inch of weld $= 0.707n/16 = 0.0442n$ sq.in. Also, capacity of weld $= 13,600(0.0442n) = 600n$ lb/lin in (108.0n N/mm).

As shown in Fig. 70, a rivet is said to be in single shear if the opposing forces tend to shear the shank along one plane and in *double shear* if they tend to shear it along two planes. The symbols R_{ss}, R_{ds}, and R_b are used here to designate the shearing capacity of a rivet in single shear, the shearing capacity of a rivet in double shear, and the bearing capacity of a rivet, respectively, expressed in pounds (newtons).

CAPACITY OF A RIVET

Determine the values of R_{ss}, R_{ds}, and R_b for a ¾-in. (19.05-mm) and ⅞-in. (22.23-mm) rivet.

Calculation Procedure:

1. Compute the cross-sectional area of the rivet
For the ¾-in. (19.05-mm) rivet, area $= A = 0.785(0.75)^2 = 0.4418$ sq.in. (2.8505 cm²). Likewise, for the ⅞-in. (22.23-mm) rivet, $A = 0.785(0.875)^2 = 0.6013$ sq.in. (3.8796 cm²).

2. Compute the single and double shearing capacity of the rivet
Let t denote the thickness, in inches (millimeters) of the connected member, as shown in Fig. 70. Multiply the stressed area by the allowable stress to determine the shearing capacity of the rivet. Thus, for the 34-in. (19.05-mm) rivet, $R_{ss} = 0.4418(15,000) = 6630$ lb (29,490.2 N); $R_{ds} = 2(0.4418)(15,000) = 13,250$ lb (58,936.0 N). Note that the factor of 2 is used for a rivet in double shear.

Likewise, for the ⅞-in. (22.23-mm) rivet, $R_{ss} = 0.6013(15,000) = 9020$ lb (40,121.0 N); $R_{ds} = 2(0.6013)(15,000) = 18,040$ lb (80,242.0 N).

3. Compute the rivet bearing capacity
The effective bearing area of a rivet of diameter d in. (mm) $= dt$. Thus, for the ¾-in. (19.05-mm) rivet, $R_b = 0.75t(48,500) = 36,380t$ lb (161,709t N). For the ⅞-in. (22.23-mm) rivet, $R_b = 0.875t(48,500) = 42,440t$ lb (188,733t N). By substituting the value of t in either relation, the numerical value of the bearing capacity could be obtained.

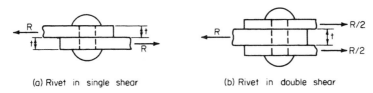

(a) Rivet in single shear   (b) Rivet in double shear

FIGURE 70

INVESTIGATION OF A LAP SPLICE

The hanger in Fig. 71a is spliced with nine ¾-in. (19.05-mm) rivets in the manner shown. Compute the load P that may be transmitted across the joint.

Calculation Procedure:

1. Compute the capacity of the joint in shear and bearing

There are three criteria to be considered: the shearing strength of the connection, the bearing strength of the connection, and the tensile strength of the net section of the plate at each row of rivets.

Since the load is concentric, assume that the load transmitted through each rivet is $1/9 P$. As plate A (Fig. 71) deflects, it bears against the upper half of each rivet. Consequently, the reaction of the rivet on plate A is exerted *above* the horizontal diametral plane of the rivet.

Computing the capacity of the joint in shear and in bearing yields $P_{ss} = 9(6630) = 59,700$ lb (265,545.6 N); $P_b = 9(36,380)(0.375) = 122,800$ lb (546,214.4 N).

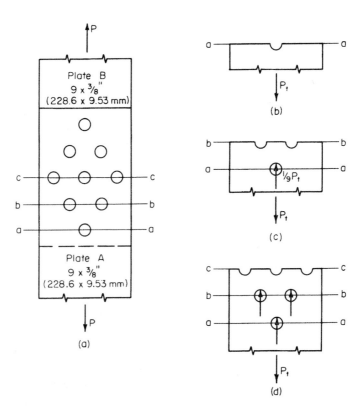

FIGURE 71

2. Compute the tensile capacity of the plate

The tensile capacity P_t lb (N) of plate A (Fig. 71) is required. In structural fabrication, rivet holes are usually punched 1/16 in. (1.59 mm) larger than the rivet diameter. However, to allow for damage to the adjacent metal caused by punching, the *effective* diameter of the hole is considered to be 1/8 in. (3.18 mm) larger than the rivet diameter.

Refer to Fig. 71b, c, and d. Equate the tensile stress at each row of rivets to 22,000 lb/sq.in. (151,690.0 kPa) to obtain P_t. Thus, at aa, residual tension = P_t net area = (9 − 0.875)(0.375) = 3.05 sq.in. (19.679 cm²). The stress $s = P_t/3.05 = 22,000$ lb/sq.in. (151,690.0 kPa); $P_t = 67,100$ lb (298,460.0 N).

At bb, residual tension = $8/9 P_t$ net area = (9 − 1.75)(0.375) = 2.72 sq.in. (17.549 cm²); $s = 8/9 P_t/2.72 = 22,000$; $P_t = 67,300$ lb (299,350.0 N).

At cc, residual tension = $2/3 P_t$ net area = (9 − 2.625)(0.375) = 2.39 sq.in. (15.420 cm²); $s = 2/3 P_t/2.39 = 22,000$; $P_t = 78,900$ lb (350,947.0 N).

3. Select the lowest of the five computed values as the allowable load

Thus, $P = 59,700$ lb (265,545.6 N).

DESIGN OF A BUTT SPLICE

A tension member in the form of a 10 × 1/2 in. (254.0 × 12.7 mm) steel plate is to be spliced with 7/8-in. (22.23-mm) rivets. Design a butt splice for the maximum load the member may carry.

Calculation Procedure:

1. Establish the design load

In a butt splice, the load is transmitted from one member to another through two auxiliary plates called *cover*, *strap*, or *splice* plates. The rivets are therefore in double shear.

Establish the design load, P lb (N), by computing the allowable load at a cross section having one rivet hole. Thus net area = (10 − 1)(0.5) = 4.5 sq.in. (29.03 cm²). Then $P = 4.5(22,000) = 99,000$ lb (440,352.0 N).

2. Determine the number of rivets required

Applying the values of rivet capacity found in an earlier calculation procedure in this section of the handbook, determine the number of rivets required. Thus, since the rivets are in double shear, $R_{ds} = 18,040$ lb (80,241.9 N); $R_b = 42,440(0.5) = 21,220$ lb (94,386.6 N). Then $99,000/18,040 = 5.5$ rivets; use the next largest whole number, or 6 rivets.

3. Select a trial pattern for the rivets; investigate the tensile stress

Conduct this investigation of the tensile stress in the main plate at each row of rivets.

The trial pattern is shown in Fig. 72. The rivet spacing satisfies the requirements of the AISC *Specification*. Record the calculations as shown:

Section	Residual tension in main plate, lb (N)	÷	Net area, sq.in. (cm²)	=	Stress, lb/in² (kPa)
aa	99,000 (440,352.0)		4.5 (29.03)		22,000 (151,690.0)
bb	82,500 (366,960.0)		4.0 (25.81)		20,600 (142,037.0)
cc	49,500 (220,176.0)		3.5 (22.58)		14,100 (97,219.5)

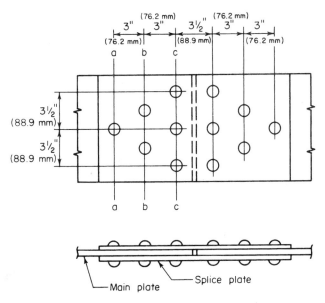

FIGURE 72

Study of the above computations shows that the rivet pattern is satisfactory.
4. Design the splice plates
To the left of the centerline, each splice plate bears against the *left* half of the rivet. Therefore, the entire load has been transmitted to the splice plates at *cc*, which is the critical section. Thus the tension in splice plate = ½(99,000) = 49,500 lb (220,176.0 N); plate thickness required = 49,500/[22,000(7)] = 0.321 in. (8.153 mm). Make the splice plates 10 × ⅜ in. (254.0 × 9.53 mm).

DESIGN OF A PIPE JOINT

A steel pipe 5 ft 6 in. (1676.4 mm) in diameter must withstand a fluid pressure of 225 lb/sq.in. (1551.4 kPa). Design the pipe and the longitudinal lap splice, using ¾-in. (19.05-mm) rivets.

Calculation Procedure:

1. Evaluate the hoop tension in the pipe
Let L denote the length (Fig. 73) of the repeating group of rivets. In this case, this equals the rivet pitch. In Fig. 73, let T denote the hoop tension, in pounds (newtons), in the distance L. Evaluate the tension, using $T = pDL/2$, where p = internal pressure, lb/sq.in. (kPa); D = inside diameter of pipe, in. (mm); L = length considered, in. (mm). Thus, $T = 225(66)L/2 = 7425L$.

2. Determine the required number of rows of rivets

Adopt, tentatively, the minimum allowable pitch, which is 2 in. (50.8 mm) for ¾-in. (19.05-mm) rivets. Then establish a feasible rivet pitch. From an earlier calculation procedure in this section, $R_{ss} = 6630$ lb (29,490.0 N). Then $T = 7425(2) = 6630n$; $n = 2.24$. Use the next largest whole number of rows, or three rows of rivets. Also, $L_{max} = 3(6630)/7425 = 2.68$ in. (68.072 mm). Use a 2½-in. (63.5-mm) pitch, as shown in Fig. 73a.

3. Determine the plate thickness

Establish the thickness t in (mm) of the steel plates by equating the stress on the net section to its allowable value. Since the holes will be drilled, take $^{13}/_{16}$ in. (20.64 mm) as their diameter. Then $T = 22,000t(2.5 - 0.81) = 7425(2.5)$; $t = 0.50$ in. (12.7 mm); use ½-in. (12.7-mm) plates. Also, $R_b = 36,380(0.5) > 6630$ lb (29,490.2 N). The rivet capacity is therefore limited by shear, as assumed.

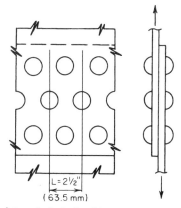

(a) Longitudinal pipe joint

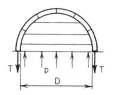

(b) Free-body diagram of upper half of pipe and contents

FIGURE 73

MOMENT ON RIVETED CONNECTION

The channel in Fig. 74a is connected to its supporting column with ¾-in. (19.05-mm) rivets and resists the couple indicated. Compute the shearing stress in each rivet.

Calculation Procedure:

1. Compute the polar moment of inertia of the rivet group

The moment causes the channel (Fig. 74) to rotate about the centroid of the rivet group and thereby exert a tangential thrust on each rivet. This thrust is directly proportional to the radial distance to the center of the rivet.

Establish coordinate axes through the centroid of the rivet group. Compute the polar moment of inertia of the group with respect to an axis through its centroid, taking the cross-sectional area of a rivet as unity. Thus, $J = \Sigma(x^2 + y^2) = 8(2.5)^2 + 4(1.5)^2 + 4(4.5)^2 = 140$ sq.in. (903.3 cm²).

2. Compute the radial distance to each rivet

Using the right-angle relationship, we see that $r_1 = r_4 = (2.5^2 + 4.5^2)^{0.5} = 5.15$ in. (130.810 mm); $r_2 = r_3 = (2.5^2 + 1.5^2)^{0.5} = 2.92$ in. (74.168 mm).

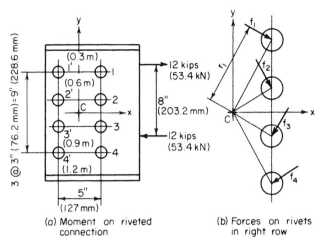

(a) Moment on riveted connection

(b) Forces on rivets in right row

FIGURE 74

3. Compute the tangential thrust on each rivet
Use the relation $f = Mr/J$. Since $M = 12{,}000(8) = 96{,}000$ lb·in. (10,846.1 N·m), $f_1 = f_4 = 96{,}000(5.15)/140 = 3530$ lb (15,701.4 N); and $f_2 = f_3 = 96{,}000(2.92)/140 = 2000$ lb (8896.0 N). The directions are shown in Fig. 58b.

4. Compute the shearing stress
Using $s = P/A$, we find $s_1 = s_4 = 3530/0.442 = 7990$ lb/sq.in. (55,090 kPa); also, $s_2 = s_3 = 2000/0.442 = 4520$ lb/sq.in. (29,300 kPa).

5. Check the rivet forces
Check the rivet forces by summing their moments with respect to an axis through the centroid. Thus $M_1 = M_4 = 3530(5.15) = 18{,}180$ in·lb (2054.0 N·m); $M_2 = M_3 = 2000(2.92) = 5840$ in·lb (659.8 N·m). Then $EM = 4(18{,}180) + 4(5840) = 96{,}080$ in·lb (10,855.1 N·m).

ECCENTRIC LOAD ON RIVETED CONNECTION

Calculate the maximum force exerted on a rivet in the connection shown in Fig. 75a.

Calculation Procedure:

1. Compute the effective eccentricity
To account implicitly for secondary effects associated with an eccentrically loaded connection, the AISC *Manual* recommends replacing the true eccentricity with an *effective eccentricity*.

To compute the effective eccentricity, use $e_e = e_a - (1 + n)/2$, where e_e = effective eccentricity, in. (mm); e_a = actual eccentricity of the load, in. (mm); n = number of rivets in a vertical row. Substituting gives $e_e = 8 - (1 + 3)/2 = 6$ in. (152.4 mm).

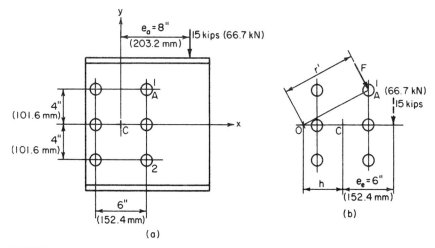

FIGURE 75

2. Replace the eccentric load with an equivalent system
The equivalent system is comprised of a concentric load P lb (N) and a clockwise moment M in·lb (N·m). Thus, $P = 15,000$ lb (66,720.0 N), $M = 15,000(6) = 90,000$ in·lb (10,168.2 N·m).

3. Compute the polar moment of inertia of the rivet group
Compute the polar moment of inertia of the rivet group with respect to an axis through its centroid. Thus, $J = \Sigma(x^2 + y^2) = 6(3)^2 + 4(4)^2 = 118$ sq.in. (761.3 cm²).

4. Resolve the tangential thrust on each rivet into its horizontal and vertical components
Resolve the tangential thrust f lb (N) on each rivet caused by the moment into its horizontal and vertical components, f_x and f_y, respectively. These forces are as follows: $f_x = My/J$ and $f_y = Mx/J$. Computing these forces for rivets 1 and 2 (Fig. 75) yields $f_x = 90,000(4)7118 = 3050$ lb (13,566.4 N); $f_y = 90,000(3)7118 = 2290$ lb (10,185.9 N).

5. Compute the thrust on each rivet caused by the concentric load
This thrust is $f'_y = 15,000/6 = 2500$ lb (11,120.0 N).

6. Combine the foregoing results to obtain the total force on the rivets being considered
The total force F lb (N) on rivets 1 and 2 is desired. Thus, $F_x = f_x = 3050$ lb (13,566.4 N); $F_y = f_y + f'_y = 2290 + 2500 = 4790$ lb (21,305.9 N). Then $F = [(3050)^2 + (4790)^2]^{0.5} = 5680$ lb (25,264.6 N).

The above six steps comprise method 1. A second way of solving this problem, method 2, is presented below.

The total force on each rivet may also be found by locating the instantaneous center of rotation associated with this eccentric load and treating the connection as if it were subjected solely to a moment (Fig. 75b).

7. Locate the instantaneous center of rotation
To locate this center, apply the relation $h = J/(e_e N)$, where N = total number of rivets and the other relations are as given earlier. Then $h = 118/[6(6)] = 3.28$ in. (83.31 m).

8. Compute the force on the rivets

Considering rivets 1 and 2, use the equation $F = Mr'/J$, where r' = distance from the instantaneous center of rotation O to the center of the given rivet, in. For rivets 1 and 2, $r' = 7.45$ in. (189.230 mm). Then $F = 90,000(7.45)/118 = 5680$ lb (25,264.6 N). The force on rivet 1 has an action line normal to the radius OA.

DESIGN OF A WELDED LAP JOINT

The 5-in. (127.0-mm) leg of a $5 \times 3 \times \frac{3}{8}$ in. ($127.0 \times 76.2 \times 9.53$ mm) angle is to be welded to a gusset plate, as shown in Fig. 76. The member will be subjected to repeated variation in stress. Design a suitable joint.

Calculation Procedure:

1. Determine the properties of the angle

In accordance with the AISC *Specification*, arrange the weld to have its centroidal axis coincide with that of the member. Refer to the AISC *Manual* to obtain the properties of the angle. Thus $A = 2.86$ sq.in. (18.453 cm^2); $y_1 = 1.70$ in. (43.2 mm); $y_2 = 5.00 - 1.70 = 3.30$ in. (83.820 mm).

2. Compute the design load and required weld length

The design load P lb (N) = $As = 2.86(22,000) = 62,920$ lb (279,868.2 N). The AISC *Specification* restricts the weld size to $\frac{5}{16}$ in. (7.94 mm). Hence, the weld capacity = 5(600) 3000 lb/lin in (525,380.4 N/m); L = weld length, in. (mm) = P/capacity, lb/lin in = $62,920/3000 = 20.97$ in. (532.638 mm).

3. Compute the joint dimensions

In Fig. 76, set $c = 5$ in. (127.0 mm), and compute a and b by applying the following equations: $a = Ly_2/w - c/2$; $b = Ly_1/w - c/2$. Thus, $a = (20.97 \times 3.30)/5 - \frac{5}{2} = 11.34$ in. (288.036 mm); $b = (20.97 \times 1.70)/5 - \frac{5}{2} = 4.63$ in. (117.602 mm). Make $a = 11.5$ in. (292.10 mm) and $b = 5$ in. (127.0 mm).

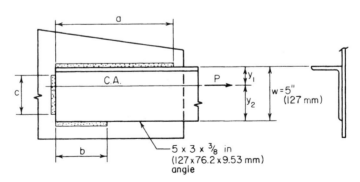

FIGURE 76

ECCENTRIC LOAD ON A WELDED CONNECTION

The bracket in Fig. 77 is connected to its support with a ¼-in. (6.35-mm) fillet weld. Determine the maximum stress in the weld.

Calculation Procedure:

1. Locate the centroid of the weld group

Refer to the previous eccentric-load calculation procedure. This situation is analogous to that. Determine the stress by locating the instantaneous center of rotation. The maximum stress occurs at A and B (Fig. 61).

Considering the weld as concentrated along the edge of the supported member, locate the centroid of the weld group by taking moments with respect to line aa. Thus $m = 2(4)(2)/(12 + 2 \times 4) = 0.8$ in. (20.32 mm).

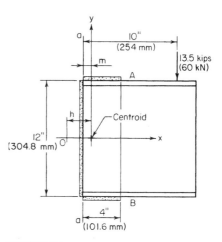

FIGURE 77

2. Replace the eccentric load with an equivalent concentric load and moment

Thus $P = 13,500$ lb (60,048.0 N); $M = 124,200$ in·lb (14,032.1 N·m).

3. Compute the polar moment of inertia of the weld group

This moment should be computed with respect to an axis through the centroid of the weld group. Thus $I_x = (1/12)(12)^3 + 2(4)(6)^2 = 432$ in³ (7080.5 cm³); $I_y = 12(0.8)^2 + 2(1/12)(4)^3 + 2(4)(2 - 0.8)^2 = 29.9$ in³ (490.06 cm³). Then $J = I_x + I_y = 461.9$ in³ (7570.54 cm³).

4. Locate the instantaneous center of rotation O

This center is associated with this eccentric load by applying the equation $h = J/(eL)$, where e = eccentricity of load, in. (mm), and L = total length of weld, in. (mm). Thus, $e = 10 - 0.8 = 9.2$ in. (233.68 mm); $L = 12 + 2(4) = 20$ in. (508.0 mm); then $h = 461.9/[9.2(20)] = 2.51$ in. (63.754 mm).

5. Compute the force on the weld

Use the equation $F = Mr'/J$, lb/lin in (N/m), where r' = distance from the instantaneous center of rotation to the given point, in. (mm). At A and B, $r' = 8.28$ in. (210.312 mm); then $F = [124,200(8.28)]/461.9 = 2230$ lb/lin in (390,532.8 N/m).

6. Calculate the corresponding stress on the throat

Thus, $s = P/A = 2230/[0.707(0.25)] = 12,600$ lb/sq.in. 86,877.0 kPa), where the value 0.707 is the sine of 45°, the throat angle.

Plastic Design of Steel Structures

Consider that a structure is subjected to a gradually increasing load until it collapses. When the yield-point stress first appears, the structure is said to be in a state of *initial yielding*. The load that exists when failure impends is termed the *ultimate load*.

In elastic design, a structure has been loaded to capacity when it attains initial yielding, on the theory that plastic deformation would annul the utility of the structure. In plastic design, on the other hand, it is recognized that a structure may be loaded beyond initial yielding if:

1. The tendency of the fiber at the yield-point stress toward plastic deformation is resisted by the adjacent fibers.
2. Those parts of the structure that remain in the elastic-stress range are capable of supporting this incremental load.

The ultimate load is reached when these conditions cease to exist and thus the structure collapses.

Thus, elastic design is concerned with an allowable *stress*, which equals the yield-point stress divided by an appropriate factor of safety. In contrast, plastic design is concerned with an allowable *load*, which equals the ultimate load divided by an appropriate factor called the *load factor*. In reality, however, the distinction between elastic and plastic design has become rather blurred because specifications that ostensibly pertain to elastic design make covert concessions to plastic behavior. Several of these are underscored in the calculation procedures that follow.

In the plastic analysis of flexural members, the following simplifying assumptions are made:

1. As the applied load is gradually increased, a state is eventually reached at which all fibers at the section of maximum moment are stressed to the yield-point stress, in either tension or compression. The section is then said to be in a state of *plastification*.
2. While plastification is proceeding at one section, the adjacent sections retain their linear-stress distribution.

Although the foregoing assumptions are fallacious, they introduce no appreciable error.

When plastification is achieved at a given section, no additional bending stress may be induced in any of its fibers, and the section is thus rendered impotent to resist any incremental bending moment. As loading continues, the beam behaves as if it had been constructed with a hinge at the given section. Consequently, the beam is said to have developed a *plastic hinge* (in contradistinction to a true hinge) at the plastified section.

The *yield moment* M_y of a beam section is the bending moment associated with initial yielding. The plastic moment M_p is the bending moment associated with plastification.

The *plastic modulus* Z of a beam section, which is analogous to the section modulus used in elastic design, is defined by $Z = M_p/f_y$, where f_y denotes the yield-point stress. The *shape factor* SF is the ratio of M_p to M_y, being so named because its value depends on the shape of the section. Then SF $= M_p/M_y = f_y Z/(f_y S) = Z/S$.

In the following calculation procedures, it is understood that the members are made of A36 steel.

ALLOWABLE LOAD ON BAR SUPPORTED BY RODS

A load is applied to a rigid bar that is symmetrically supported by three steel rods as shown in Fig. 78. The cross-sectional areas of the rods are: rods A and C, 1.2 sq.in. (7.74 cm²); rod B, 1.0 sq.in. (6.45 cm²). Determine the maximum load that may be applied, (*a*) using elastic design with an allowable stress of 22,000 lb/sq.in. (151,690.0 kPa); (b) using plastic design with a load factor of 1.85.

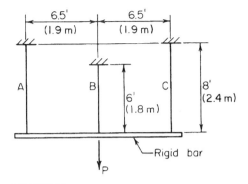

FIGURE 78

Calculation Procedure:

1. Express the relationships among the tensile stresses in the rods

The symmetric disposition causes the bar to deflect vertically without rotating, thereby elongating the three rods by the same amount. As the first method of solving this problem, assume that the load is gradually increased from zero to its allowable value.

Expressing the relationships among the tensile stresses, we have $\Delta L = s_A L_A/E = s_B L_B/E = s_C L_C/E$; therefore, $s_A = s_C$, and $s_A = s_B L_B/L_A = 0.75 s_B$ for this arrangement of rods. Since s_B is the maximum stress, the allowable stress first appears in rod B.

2. Evaluate the stresses at the instant the load attains its allowable value

Calculate the load carried by each rod, and sum these loads to find P_{allow}. Thus $s_B = 22,000$ lb/sq.in. (151,690.0 kPa); $s_B = 0.75(22,000) = 16,500$ lb/sq.in. (113,767.5 kPa); $P_A = P_C = 16,500(1.2) = 19,800$ lb (88,070.4 N); $P_B = 22,000(1.0) = 22,000$ lb (97,856.0 N); $P_{\text{allow}} = 2(19,800) + 22,000 = 61,600$ lb (273,996.8 N).

Next, consider that the load is gradually increased from zero to its ultimate value. When rod B attains its yield-point stress, its tendency to deform plastically is inhibited by rods A and C because the rigidity of the bar constrains the three rods to elongate uniformly. The structure therefore remains stable as the load is increased beyond the elastic range until rods A and C also attain their yield-point stress.

3. Find the ultimate load

To find the ultimate load P_u, equate the stress in each rod to f_y, calculate the load carried by each rod, and sum these loads to find the ultimate load P_u. Thus, $P_A = P_C = 36,000(1.2) = 43,200$ lb (192,153.6 N); $P_B = 36,000(1.0) = 36,000$ lb (160,128.0 N); $P_u = 2(43,200) + 36,000 = 122,400$ lb (544,435.2 N).

4. Apply the load factor to establish the allowable load

Thus, $P_{\text{allow}} = P_u/\text{LF} = 122,400/1.85 = 66,200$ lb (294,457.6 N).

DETERMINATION OF SECTION SHAPE FACTORS

Without applying the equations and numerical values of the plastic modulus given in the AISC *Manual*, determine the shape factor associated with a rectangle, a circle, and a W16 × 40. Explain why the circle has the highest and the W section the lowest factor of the 3.

Calculation Procedure:

1. Calculate M_y for each section

Use the equation $M_y = Sf_y$ for each section. Thus, for a rectangle, $M_y = bd^2 f_y/6$. For a circle, using the properties of a circle as given in the *Manual*, we find $M_y = \pi d^3 f_y/32$. For a W16 × 40,

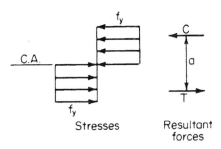

FIGURE 79. Conditions at section of plastification.

$A = 11.77$ sq.in. (75.940 cm²), $S = 64.4$ in³ (1055.52 cm³), and $M_y = 64.4 f_y$.

2. Compute the resultant forces associated with plastification
In Fig. 79, the resultant forces are C and T. Once these forces are known, their action lines and M_p should be computed.

Thus, for a rectangle, $C = bd f_y/2$, $a = d/2$, and $M_p = aC = bd^2 f_y/4$. For a circle, $C = \pi d^2 f_y/8$, $a = 4d/(3\pi)$, and $M_p = aC = d^3 f_y/6$. For a W16 × 40, $C = \frac{1}{2}(11.77$ sq.in.$) = 5.885 f_y$.

To locate the action lines, refer to the Manual and note the position of the centroidal axis of the WT8 × 20 section, i.e., a section half the size of that being considered. Thus, $a = 2(8.00 - 1.82) = 12.36$ in. (313.944 mm); $M_p = aC = 12.36(5.885 f_y) = 72.7 f_y$.

3. Divide M_p by M_y to obtain the shape factor
For a rectangle, SF $= (bd^2/4)/(bd^2/6) = 1.50$. For a circle, SF $= (d^3/6)/(\pi d^3/32) = 1.70$. For a WT16 × 40, SF $= 72.7/64.4 = 1.13$.

4. Explain the relative values of the shape factor
To explain the relative values of the shape factor, express the resisting moment contributed by a given fiber at plastification and at initial yielding, and compare the results. Let dA denote the area of the given fiber and y its distance from the neutral axis. At plastification, $dM_p = f_y y dA$. At initial yielding, $f = f_y y/c$; $dM_y = f_y y^2 dA/c$; $dM_p/dM_y = c/y$.

By comparing a circle and a hypothetical W section having the same area and depth, the circle is found to have a larger shape factor because of its relatively low values of y.

As this analysis demonstrates, the process of plastification mitigates the detriment that accrues from placing any area near the neutral axis, since the stress at plastification is independent of the position of the fiber. Consequently, a section that is relatively inefficient with respect to flexure has a relatively high shape factor. The AISC *Specification* for elastic design implicitly recognizes the value of the shape factor by assigning an allowable bending stress of $0.75 f_y$ to rectangular bearing plates and $0.90 f_y$ to pins.

DETERMINATION OF ULTIMATE LOAD BY THE STATIC METHOD

The W18 × 45 beam in Fig. 80a is simply supported at A and fixed at C. Disregarding the beam weight, calculate the ultimate load that may be applied at B (a) by analyzing the behavior of the beam during its two phases; (b) by analyzing the bending moments that exist at impending collapse. (The first part of the solution illustrates the postelastic behavior of the member.)

Calculation Procedure:

1. Calculate the ultimate-moment capacity of the member
Part a: As the load is gradually increased from zero to its ultimate value, the beam passes through two phases. During phase 1, the *elastic phase*, the member is restrained against

rotation at C. This phase terminates when a plastic hinge forms at that end. During phase 2-the *postelastic*, or *plastic, phase*—the member functions as a simply supported beam. This phase terminates when a plastic hinge forms at B, since the member then becomes unstable.

Using data from the AISC *Manual*, we have $Z = 89.6$ in^3 (1468.54 cm^3). Then $M_p = f_y Z = 36(89.6)/12 = 268.8$ ft·kips (364.49 kN·m).

2. Calculate the moment BD

Let P_1 denote the applied load at completion of phase 1. In Fig. 80b, construct the bending-moment diagram ADEC corresponding to this load. Evaluate P_1 by applying the equations for case 14 in the AISC *Manual*. Calculate the moment BD. Thus, $CE = -ab(a + L)P_1/(2L^2) = -20(10)(50)P_1/[2(900)] = -268.8$; $P_1 = 48.38$ kips (215.194 kN); $BD = ab^2(a + 2L)P_1/(2L^3) = 20(100)(80)(48.38)/[2(27,000)] = 143.3$ ft·kips (194.31 kN·m).

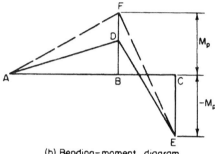

(a) Force diagram

(b) Bending-moment diagram

FIGURE 80

3. Determine the incremental load at completion of phase 2

Let P_2 denote the incremental applied load at completion of phase 2, i.e., the actual load on the beam minus P_1. In Fig. 80b, construct the bending-moment diagram AFEC that exists when phase 2 terminates. Evaluate P_2 by considering the beam as simply supported. Thus, $BF = 268.8$ ft·kips (364.49 kN·m); $DF = 268.8 - 143.3 = 125.5$ ft·kips (170.18 kN·m); but $DF = abP_2/L = 20(10)P_2/30 = 125.5$; $P_2 = 18.82$ kips (83.711 kN).

4. Sum the results to obtain the ultimate load

Thus, $P_u = 48.38 + 18.82 = 67.20$ kips (298.906 kN).

5. Construct the force and bending-moment diagrams for the ultimate load

Part b: The following considerations are crucial: The bending-moment diagram always has vertices at B and C, and formation of two plastic hinges will cause failure of the beam. Therefore, the plastic moment occurs at B and C at impending failure. *The sequence in which the plastic hinges are formed at these sections is immaterial.*

These diagrams are shown in Fig. 81. Express M_p in terms of P_u, and evaluate P_u. Thus, $BF = 20R_A = 268.8$; therefore, $R_A = 13.44$ kips (59.781 kN). Also, $CE = 30R_A - 10P_u = 30 \times 13.44 - 10P_u = -268.8$; $P_u = 67.20$ kips (298.906 kN).

Here is an alternative method: $BF = (abP_u/L) - aM_p/L = M_p$, or $20(10)P_u/30 = 50M_p/30$; $P_u = 67.20$ kips (298.906 kN).

This solution method used in part b is termed the static, or equilibrium, method. As this solution demonstrates, it is unnecessary to trace the stress history of the member as it passes through its successive phases, as was done in part *a*; the analysis can be confined to

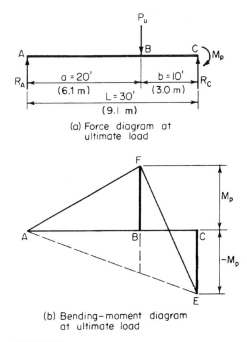

(a) Force diagram at ultimate load

(b) Bending–moment diagram at ultimate load

FIGURE 81

the conditions that exist at impending failure. This procedure also illustrates the following important characteristics of plastic design:

1. Plastic design is far simpler than elastic design.
2. Plastic design yields results that are much more reliable than those secured through elastic design. For example, assume that the support at C does not completely inhibit rotation at that end. This departure from design conditions will invalidate the elastic analysis but will in no way affect the plastic analysis.

DETERMINING THE ULTIMATE LOAD BY THE MECHANISM METHOD

Use the mechanism method to solve the problem given in the previous calculation procedure.

Calculation Procedure:

1. Indicate, in hyperbolic manner, the virtual displacement of the member from its initial to a subsequent position

To the two phases of beam behavior previously considered, it is possible to add a third. Consider that when the ultimate load is reached, the member is subjected to an incremental deflection. This will result in collapse, but the behavior of the member can be analyzed during an infinitesimally small deflection from its stable position. This is termed a virtual deflection, or displacement.

Since the member is incapable of supporting any load beyond that existing at completion of phase 2, this virtual deflection is not characterized by any change in bending stress. Rotation therefore occurs solely at the real and plastic hinges. Thus, during phase 3, the member behaves as a mechanism (i.e., a constrained chain of pin-connected rigid bodies, or links).

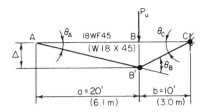

FIGURE 82

In Fig. 82, indicate, in hyperbolic manner, the virtual displacement of the member from its initial position ABC to a subsequent position $AB'C$. Use dots to represent plastic hinges. (The initial position may be represented by a straight line for simplicity because the analysis is concerned solely with the deformation that occurs *during* phase 3.)

2. Express the linear displacement under the load and the angular displacement at every plastic hinge

Use a convenient unit to express these displacements. Thus, $\Delta = a\theta_A = b\theta_C$; therefore, $\theta_C = a\theta_A/b = 2\theta_A$; $\theta_B = \theta_A + \theta_C = 3\theta_A$.

3. Evaluate the external and internal work associated with the virtual displacement

The work performed by a constant force equals the product of the force and its displacement parallel to its action line. Also, the work performed by a constant moment equals the product of the moment and its angular displacement. Work is a positive quantity when the displacement occurs in the direction of the force or moment. Thus, the external work $W_E = P_u\Delta = P_u a\theta_A = 20P_u\theta_A$. And the internal work $W_I = M_p(\theta_B + \theta_C) = 5M_p\theta_A$.

4. Equate the external and internal work to evaluate the ultimate load

Thus, $20P_u\theta_A = 5M_p\theta_A$; $P_u = (5/20)(268.8) = 67.20$ kips (298.906 kN).

The solution method used here is also termed the *virtual-work*, or *kinematic*, method.

ANALYSIS OF A FIXED-END BEAM UNDER CONCENTRATED LOAD

If the beam in the two previous calculation procedures is fixed at A as well as at C, what is the ultimate load that may be applied at B?

Calculation Procedure:

1. Determine when failure impends

When hinges form at A, B, and C, failure impends. Repeat steps 3 and 4 of the previous calculation procedure, modifying the calculations to reflect the revised conditions. Thus $W_E = 20P_u\theta_A$; $W_I = M_P(\theta_a + \theta_B + \theta_C) = 6M_p\theta_A$; $20P_u\theta_A = 6M_p\theta_A$; $P_u = (6/20)(268.8) = 80.64$ kips (358.687 kN).

2. Analyze the phases through which the member passes

This member passes through three phases until the ultimate load is reached. Initially, it behaves as a beam fixed at both ends, then as a beam fixed at the left end only, and finally as a simply supported beam. However, as already discussed, these considerations are extraneous in plastic design.

ANALYSIS OF A TWO-SPAN BEAM WITH CONCENTRATED LOADS

The continuous W18 × 45 beam in Fig. 83 carries two equal concentrated loads having the locations indicated. Disregarding the weight of the beam, compute the ultimate value of these loads, using both the static and the mechanism method.

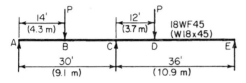

FIGURE 83

Calculation Procedure:

1. Construct the force and bending-moment diagrams

The continuous beam becomes unstable when a plastic hinge forms at C and at another section. The bending-moment diagram has vertices at B and D, but it is not readily apparent at which of these sections the second hinge will form. The answer is found by assuming a plastic hinge at B and at D, in turn, computing the corresponding value of P_u, and selecting the lesser value as the correct result. Part a will use the static method; part b, the mechanism method.

Assume, for part a, a plastic hinge at B and C. In Fig. 84, construct the force diagram and bending-moment diagram for span AC. The moment diagram may be drawn in the manner

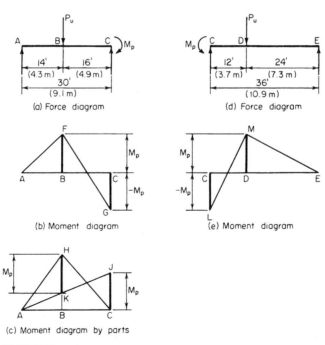

FIGURE 84

shown in Fig. 84b or c, whichever is preferred. In Fig. 84c, ACH represents the moments that would exist in the absence of restraint at C, and ACJ represents, in absolute value, the moments induced by this restraint. Compute the load P_u associated with the assumed hinge location. From previous calculation procedures, $M_p = 268.8$ ft·kips (364.49 kN·m); then $M_B = 14 \times 16P_u/30 - 14M_p/30 = M_p$; $P_u = 44(268.8)/224 = 52.8$ kips (234.85 kN).

2. Assume another hinge location and compute the ultimate load associated with this location

Now assume a plastic hinge at C and D. In Fig. 84, construct the force diagram and bending-moment diagram for CE. Computing the load P_u associated with this assumed location, we find $M_D = 12 \times 24P_u/36 - 24M_p/36 = M_p$; $P_u = 60(268.8)/288 = 56.0$ kips (249.09 kN).

3. Select the lesser value of the ultimate load

The correct result is the lesser of these alternative values, or $P_u = 52.8$ kips (234.85 kN). At this load, plastic hinges exist at B and C but not at D.

4. For the mechanism method, assume a plastic-hinge location

It will be assumed that plastic hinges are located at B and C (Fig. 85). Evaluate P_u. Thus, $\theta_C = 14\theta_A/16$; $\theta_B = 30\theta_A/16$; $\Delta = 14\theta_A$; $W_E = P_u\Delta = 14P_u\theta_A$; $W_I = M_p(\theta_B + \theta_C) = 2.75M_p\theta_A$; $14P_u\theta_A = 2.75M_p\theta_A$; $P_u = 52.8$ kips (234.85 kN).

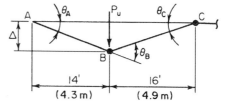

FIGURE 85

5. Assume a plastic hinge at another location

Select C and D for the new location. Repeat the above procedure. The result will be identical with that in step 2.

SELECTION OF SIZES FOR A CONTINUOUS BEAM

Using a load factor of 1.70, design the member to carry the working loads (with beam weight included) shown in Fig. 86a. The maximum length that can be transported is 60 ft (18.3 m).

Calculation Procedure:

1. Determine the ultimate loads to be supported

Since the member must be spliced, it will be economical to adopt the following design:
a. Use the particular beam size required for each portion, considering that the two portions will fail simultaneously at ultimate load. Therefore, three plastic hinges will exist at failure—one at the interior support and one in the interior of each span.
b. Extend one beam beyond the interior support, splicing the member at the point of contraflexure in the adjacent span. Since the maximum simple-span moment is greater for AB than for BC, it is logical to assume that for economy the left beam rather than the right one should overhang the support.

Multiply the working loads by the load factor to obtain the ultimate loads to be supported. Thus, $w = 1.2$ kips/lin ft (17.51 kN/m); $w_u = 1.70(1.2) = 2.04$ kips/lin ft (29.77 kN/m); $P = 10$ kips (44.5 kN); $P_u = 1.70(10) = 17$ kips (75.6 kN).

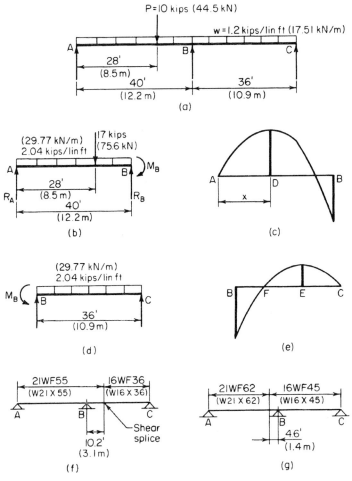

FIGURE 86

2. Construct the ultimate-load and corresponding bending-moment diagram for each span

Set the maximum positive moment M_D in span AB and the negative moment at B equal to each other in absolute value.

3. Evaluate the maximum positive moment in the left span

Thus, $R_A = 45.9 - M_B/40$; $x = R_A/2.04$; $M_D = \frac{1}{2}R_A x = R_A^2/4.08 = M_B$. Substitute the value of R_A and solve. Thus, $M_D = 342$ ft·kips (463.8 kN·m).

An indirect but less cumbersome method consists of assigning a series of trial values to M_B and calculating the corresponding value of M_D, continuing the process until the required equality is obtained.

4. Select a section to resist the plastic moment

Thus, $Z = M_p/f_y = 342(12)136 = 114$ in^3 (1868.5 cm^3). Referring to the AISC *Manual*, use a W21 × 55 with $Z = 125.4$ in^3 (2055.31 cm^3).

5. Evaluate the maximum positive moment in the right span

Equate M_B to the true plastic-moment capacity of the W21 × 55. Evaluate the maximum positive moment M_E in span BC, and locate the point of contraflexure. Therefore, $M_B = -36(125.4)/12 = -376.2$ ft·kips (−510.13 kN·m); $M_E = 169.1$ ft·kips (229.30 kN·m); $BF = 10.2$ ft (3.11 m).

6. Select a section to resist the plastic moment

The moment to be resisted is M_E. Thus, $Z = 169.1(12)/36 = 56.4$ in³ (924.40 cm³). Use W16 × 36 with $Z = 63.9$ in³ (1047.32 cm³).

The design is summarized in Fig. 86f. By inserting a hinge at F, the continuity of the member is destroyed and its behavior is thereby modified under gradually increasing load. However, the ultimate-load conditions, which constitute the only valid design criteria, are not affected.

7. Alternatively, design the member with the right-hand beam overhanging the support

Compare the two designs for economy. The latter design is summarized in Fig. 86g. The total beam weight associated with each scheme is as shown in the following table.

Design 1	Design 2
55(50.2) = 2,761 lb (12,280.9 N)	62(35.4) = 2,195 lb (9,763.4 N)
36(25.8) = 929 lb (4,132.2 N)	45(40.6) = 1,827 lb (8,126.5 N)
Total 3,690 lb (16,413.1 N)	4,022 lb (17,889.9 N)

For completeness, the column sizes associated with the two schemes should also be compared.

MECHANISM-METHOD ANALYSIS OF A RECTANGULAR PORTAL FRAME

Calculate the plastic moment and the reactions at the supports at ultimate load of the prismatic frame in Fig. 87a. Use a load factor of 1.85, and apply the mechanism method.

Calculation Procedure:

1. Compute the ultimate loads to be resisted

There are three potential modes of failure to consider:

a. Failure of the beam BD through the formation of plastic hinges at B, C, and D (Fig. 87b)
b. Failure by sidesway through the formation of plastic hinges at B and D (Fig. 87c)
c. A composite of the foregoing modes of failure, characterized by the formation of plastic hinges at C and D

Since the true mode of failure is not readily discernible, it is necessary to analyze each of the foregoing. The true mode of failure is the one that yields the highest value of M_p.

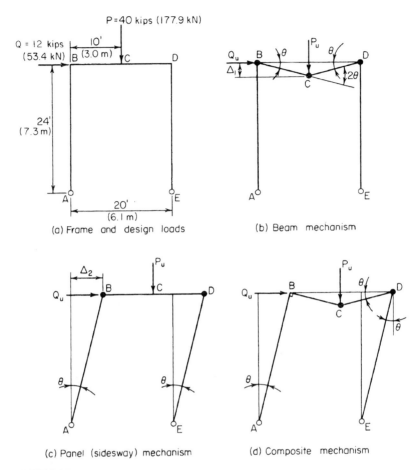

(a) Frame and design loads
(b) Beam mechanism
(c) Panel (sidesway) mechanism
(d) Composite mechanism

FIGURE 87

Although the work quantities are positive, it is advantageous to supply each angular displacement with an algebraic sign. A rotation is considered positive if the angle on the interior side of the frame increases. The algebraic sum of the angular displacements must equal zero.

Computing the ultimate loads to be resisted yields $P_u = 1.85(40) = 74$ kips (329.2 kN); $Q_u = 1.85(12) = 22.2$ kips (98.75 kN).

2. Assume the mode of failure in Fig. 87b and compute M_p

Thus, $\Delta_1 = 10\theta$; $W_E = 74(10\theta) = 740\theta$. Then indicate in a tabulation, such as that shown here, where the plastic moment occurs. Include all significant sections for completeness.

Section	Angular displacement	Moment	W_1
A			
B	$-\theta$	M_p	$M_p\theta$
C	$+2\theta$	M_p	$2M_p\theta$
D	$-\theta$	M_p	$M_p\theta$
E	. .	. .	. .
Total			$4M_p\theta$

Then $4M_p\theta = 740\theta$; $M_p = 185$ ft·kips (250.9 kN·m).

3. Repeat the foregoing procedure for failure by sidesway

Thus, $\Delta_2 = 24\theta$; $W_E = 22.2(24\theta) = 532.8\theta$.

Section	Angular displacement	Moment	W_1
A	$-\theta$		
B	$+\theta$	M_p	$M_p\theta$
C			
D	$-\theta$	M_p	$M_p\theta$
E	$+\theta$		
Total			$2M_p\theta$

Then $2M_p\theta = 532.8\theta$; $M_p = 266.4$ ft·kips (361.24 kN·m).

4. Assume the composite mode of failure and compute M_p

Since this results from superposition of the two preceding modes, the angular displacements and the external work may be obtained by adding the algebraic values previously found. Thus, $W_E = 740\theta + 532.8\theta = 1272.8\theta$. Then the tabulation is as shown:

Section	Angular displacement	Moment	W_1
A	$-\theta$		
B			
C	$+2\theta$	M_p	$2M_p\theta$
D	-2θ	M_p	$2M_p\theta$
E	$+\theta$		
Total			$4M_p\theta$

Then $4M_p\theta = 1272.8\theta$; $M_p = 318.2$ ft·kips (431.48 kN·m).

5. Select the highest value of M_p as the correct result

Thus, $M_p = 318.2$ ft·kips (431.48 kN·m). The structure fails through the formation of plastic hinges at C and D. That a hinge should appear at D rather than at B is plausible when it is considered that the bending moments induced by the two loads are of like sign at D but of opposite sign at B.

6. Compute the reactions at the supports

Draw a free-body diagram of the frame at ultimate load (Fig. 88). Compute the reactions at the supports by applying the computed values of M_C and M_D. Thus, $\Sigma M_E = 20V_A + 22.2(24) - 74(10) = 0$; $V_A = 10.36$ kips (46.081 kN); $V_E = 74 - 10.36 = 63.64$ kips (283.071 kN); $M_C = 10V_A + 24H_A = 103.6 + 24H_A = 318.2$; $H_A = 8.94$ kips (39.765 kN); $H_E = 22.2 - 8.94 = 13.26$ kips (58.980 kN); $M_D = -24H_E = -24(13.26) = -318.2$ ft·kips (−431.48 kN·m). Thus, the results are verified.

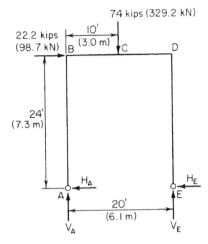

FIGURE 88

ANALYSIS OF A RECTANGULAR PORTAL FRAME BY THE STATIC METHOD

Compute the plastic moment of the frame in Fig. 87a by using the static method.

Calculation Procedure:

1. Determine the relative values of the bending moments

Consider a bending moment as positive if the fibers on the interior side of the neutral plane are in tension. Consequently, as the mechanisms in Fig. 87 reveal, the algebraic sign of the plastic moment at a given section agrees with that of its angular displacement during collapse.

Determine the relative values of the bending moments at B, C, and D. Refer to Fig. 88. As previously found by statics, $V_A = 10.36$ kips (46.081 kN), $M_B = 24H_A$, $M_C = 24H_A + 10V_A$; therefore, $M_C = M_B + 103.6$, Eq. a. Also, $M_D = 24H_A + 20V_A - 74(10)$; $M_D = M_B - 532.8$, Eq. b; or $M_D = M_C - 636.4$, Eq. c.

2. Assume the mode of failure in Fig. 87b

This requires that $M_B = M_D = -M_p$. This relationship is incompatible with Eq. b, and the assumed mode of failure is therefore incorrect.

3. Assume the mode of failure in Fig. 87c

This requires that $M_B = M_p$, and $M_C < M_p$; therefore, $M_C < M_B$. This relationship is incompatible with Eq. a, and the assumed mode of failure is therefore incorrect.

By a process of elimination, it has been ascertained that the frame will fail in the manner shown in Fig. 87d.

4. Compute the value of M_p for the composite mode of failure

Thus, $M_C = M_p$, and $M_D = -M_p$. Substitute these values in Eq. c. Or, $-M_p = M_p - 636.4$; $M_p = 318.2$ ft·kips (431.48 kN·m).

THEOREM OF COMPOSITE MECHANISMS

By analyzing the calculations in the calculation procedure before the last one, establish a criterion to determine when a composite mechanism is significant (i.e., under what conditions it may yield an M_p value greater than that associated with the basic mechanisms).

Calculation Procedure:

1. Express the external and internal work associated with a given mechanism

Thus, $W_E = e\theta$, and $W_I = iM_p\theta$, where the coefficients e and i are obtained by applying the mechanism method. Then $M_p = e/i$.

2. Determine the significance of mechanism sign

Let the subscripts 1 and 2 refer to the basic mechanisms and the subscript 3 to their composite mechanism. Then $M_{p1} = e_1/i_1$; $M_{p2} = e_2/i_2$.

When the basic mechanisms are superposed, the values of W_E are additive. If the two mechanisms do not produce rotations of opposite sign at any section, the values of W_I are also additive, and $M_{p3} = e_3/i_3 = (e_1 + e_2)/(i_1 + i_2)$. This value is intermediate between M_{p1} and M_{p2}, and the composite mechanism therefore lacks significance. But if the basic mechanisms produce rotations of opposite sign at any section whatsoever, M_{p3} may exceed both M_{p1} and M_{p2}.

In summary, a composite mechanism is significant only if the two basic mechanisms of which it is composed produce rotations of opposite sign at any section. This theorem, which establishes a necessary but not sufficient condition, simplifies the analysis of a complex frame by enabling the engineer to discard the nonsignificant composite mechanisms at the outset.

ANALYSIS OF AN UNSYMMETRIC RECTANGULAR PORTAL FRAME

The frame in Fig. 89a sustains the ultimate loads shown. Compute the plastic moment and ultimate-load reactions.

Calculation Procedure:

1. Determine the solution method to use

Apply the mechanism method. In Fig. 89b, indicate the basic mechanisms.

2. Identify the significant composite mechanisms

Apply the theorem of the previous calculation procedure. Using this theorem, identify the significant composite mechanisms. For mechanisms 1 and 2, the rotations at B are of opposite sign; their composite therefore warrants investigation.

For mechanisms 1 and 3, there are no rotations of opposite sign; their composite therefore fails the test. For mechanisms 2 and 3, the rotations at B are of opposite sign; their composite therefore warrants investigation.

3. Evaluate the external work associated with each mechanism

Mechanism	W_E
1	$80\Delta_1 = 80(10\theta) = 800\theta$
2	$2\Delta_2 = 20(15\theta) = 300\theta$
3	300θ
4	1100θ
5	600θ

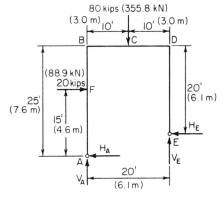

(a) Frame and ultimate loads

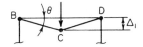

Mechanism 1

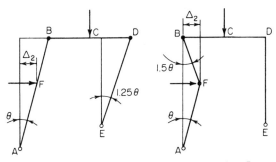

Mechanism 2　　　　　　Mechanism 3

(b) Basic mechanisms

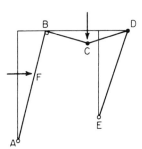

Mechanism 4　　　　　　Mechanism 5
Composite of 1 and 2　　Composite of 2 and 3

(c) Composite mechanisms

FIGURE 89

4. List the sections at which plastic hinges form; record the angular displacement associated with each mechanism

Use a list such as the following:

		Section		
Mechanism	B	C	D	F
1	$-\theta$	$+2\theta$	$-\theta$	
2	$+\theta$		-1.2θ	
3	-1.5θ			$+2.5\theta$
4		$+2\theta$	-2.25θ	
5	-0.5θ		-1.25θ	$+2.5\theta$

5. Evaluate the internal work associated with each mechanism

Equate the external and internal work to find M_p. Thus, $M_{p1} = 800/4 = 200$; $M_{p2} = 300/2.25 = 133.3$; $M_{p3} = 300/4 = 75$; $M_{p4} = 1100/4.25 = 258.8$; $M_{p5} = 600/4.25 = 141.2$. Equate the external and internal work to find M_p.

6. Select the highest value as the correct result

Thus, $M_p = 258.8$ ft·kips (350.93 kN·m). The frame fails through the formation of plastic hinges at C and D.

7. Determine the reactions at ultimate load

To verify the foregoing solution, ascertain that the bending moment does not exceed M_p in absolute value anywhere in the frame. Refer to Fig. 89a.

Thus, $M_D = -20H_E = -258.8$; therefore, $H_E = 12.94$ kips (57.557 kN); $M_C = M_D + 10V_E = 258.8$; therefore, $V_E = 51.76$ kips (230.23 kN); then $H_A = 7.06$ kips (31.403 kN); $V_A = 28.24$ kips (125.612 kN).

Check the moments. Thus $\Sigma M_E = 20V_A + 5H_A + 20(10) - 80(10) = 0$; this is correct. Also, $M_F = 15H_A = 105.9$ ft·kips (143.60 kN·m) $< M_p$. This is correct. Last, $M_B = 25H_A - 20(10) = -23.5$ ft·kips (−31.87 kN·m) $> -M_p$. This is correct.

ANALYSIS OF GABLE FRAME BY STATIC METHOD

The prismatic frame in Fig. 90a carries the ultimate loads shown. Determine the plastic moment by applying the static method.

Calculation Procedure:

1. Compute the vertical shear V_A and the bending moment at every significant section, assuming $H_A = 0$

Thus, $V_A = 41$ kips (182.4 kN). Then $M_B = 0$; $M_C = 386$; $M_D = 432$; $M_E = 276$; $M_F = -100$.

Note that failure of the frame will result from the formation of two plastic hinges. It is helpful, therefore, to construct a "projected" bending-moment diagram as an aid in locating these hinges. The computed bending moments are used in plotting the projected bending-moment diagram.

1.138　　　　　　　　STRUCTURAL ENGINEERING

2. Construct a projected bending-moment diagram
To construct this diagram, consider the rafter BD to be projected onto the plane of column AB and the rafter FD to be projected onto the plane of column GF. Juxtapose the two halves, as shown in Fig. 90b. Plot the values calculated in step 1 to obtain the bending-moment diagram corresponding to the assumed condition of $H_A = 0$.

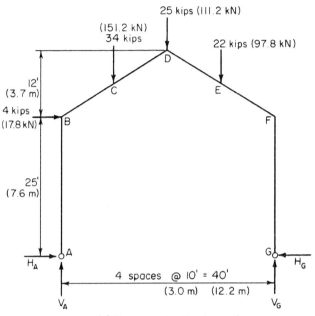

(a) Frame and ultimate loads

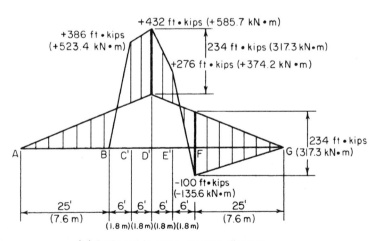

(b) Projected bending-moment diagram

FIGURE 90

The bending moments caused solely by a specific value of H_A are represented by an isosceles triangle with its vertex at D'. The true bending moments are obtained by superposition. It is evident by inspection of the diagram that plastic hinges form at D and F and that H_A is directed to the right.

3. Evaluate the plastic moment
Apply the true moments at D and F. Thus, $M_D = M_p$ and $M_F = -M_p$; therefore, $432 - 37H_A = -(-100 - 25H_A)$; $H_A = 5.35$ kips (23.797 kN) and $M_p = 234$ ft·kips (317 kN·m).

THEOREM OF VIRTUAL DISPLACEMENTS

In Fig. 91a, point P is displaced along a virtual (infinitesimally small) circular arc PP' centered at O and having a central angle θ. Derive expressions for the horizontal and vertical displacement of P in terms of the given data. (These expressions are applied later in analyzing a gable frame by the mechanism method.)

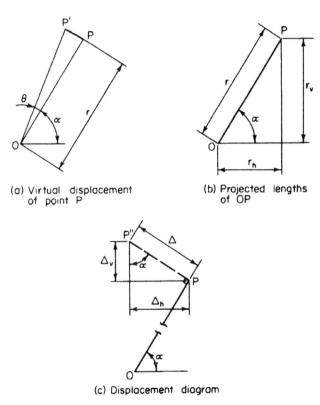

(a) Virtual displacement of point P

(b) Projected lengths of OP

(c) Displacement diagram

FIGURE 91

Calculation Procedure:

1. Construct the displacement diagram
In Fig. 91b, let r_h = length of horizontal projection of OP; r_v = length of vertical projection of OP; Δ_h = horizontal displacement of P; Δ_v = vertical displacement of P.

In Fig. 91c, construct the displacement diagram. Since PP′ is infinitesimally small, replace this circular arc with the straight line PP″ that is tangent to the arc at P and therefore normal to radius OP.

2. Evaluate Δ_h and Δ_v, considering only absolute values
Since θ is infinitesimally small, set $PP'' = r\theta$; $\Delta_h = PP''\sin\alpha = r\theta\sin\alpha$; $\Delta_v = PP''\cos\alpha = r\theta\cos\alpha$. But $r\sin\alpha = r_v$ and $r\cos\alpha = r_h$; therefore, $\Delta_h = r_v\theta$ and $\Delta_v = r_h\theta$.

These results may be combined and expressed verbally thus: If a point is displaced along a virtual circular arc, its displacement as projected on the u axis equals the displacement angle times the length of the radius as projected on an axis normal to u.

GABLE-FRAME ANALYSIS BY USING THE MECHANISM METHOD

For the frame in Fig. 90a, assume that plastic hinges form at D and F. Calculate the plastic moment associated with this assumed mode of failure by applying the mechanism method.

Calculation Procedure:

1. Indicate the frame configuration following a virtual displacement
During collapse, the frame consists of three rigid bodies: ABD, DF, and GF. To evaluate the external and internal work performed during a virtual displacement, it is necessary to locate the instantaneous center of rotation of each body.

In Fig. 92 indicate by dash lines the configuration of the frame following a virtual displacement. In Fig. 92, D is displaced to D′ and F to F′. Draw a straight line through A and D intersecting the prolongation of GF at H.

Since A is the center of rotation of ABD, DD′ is normal to AD and HD; since G is the center of rotation of GF, FF′ is normal to GF and HF. Therefore, H is the instantaneous center of rotation of DF.

2. Record the pertinent dimensions and rotations
Record the dimensions a, b, and c in Fig. 33, and express θ_2 and θ_3 in terms of θ_1. Thus, $\theta_2/\theta_1 = HD/AD$; $\therefore \theta_2 = \theta_1$. Also, $\theta_3/\theta_1 = HF/GF = 49/25$; $\therefore \theta_3 = 1.96\theta_1$.

3. Determine the angular displacement, and evaluate the internal work
Determine the angular displacement (in absolute value) at D and F, and evaluate the internal work in terms of θ_1. Thus, $\theta_D = \theta_1 + \theta_2 = 2\theta_1$; $\theta_F = \theta_1 + \theta_3 = 2.96\theta_1$. Then $W_I = M_p(\theta_D + \theta_F) = 4.96 M_p\theta_1$.

4. Apply the theorem of virtual displacements to determine the displacement of each applied load
Determine the displacement of each applied load in the direction of the load. Multiply the displacement by the load to obtain the external work. Record the results as shown:

STRUCTURAL STEEL DESIGN

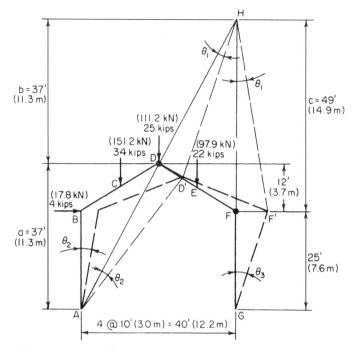

FIGURE 92. Virtual displacement of frame.

	Load		Displacement in direction of load		External work	
Section	kips	kN	ft	m	ft·kips	kN·m
B	4	17.8	$\Delta_h = 25\theta_2 = 25\theta_1$	$7.6\theta_1$	$100\theta_1$	$135.6\theta_1$
C	34	151.2	$\Delta_v = 10\theta_2 = 10\theta_1$	$3.0\theta_1$	$340\theta_1$	$461.0\theta_1$
D	25	111.2	$\Delta_v = 20\theta_1$	$6.1\theta_1$	$500\theta_1$	$678.0\theta_1$
E	22	97.9	$\Delta_v = 10\theta_1$	$3.0\theta_1$	$220\theta_1$	$298.3\theta_1$
Total					$1160\theta_1$	$1572.9\theta_1$

5. Equate the external and internal work to find M_p

Thus, $4.96 M_p \theta_1 = 1160 \theta_1$; $M_p = 234$ ft·kips (317.3 kN·m).

Other modes of failure may be assumed and the corresponding value of M_p computed in the same manner. The failure mechanism analyzed in this procedure (plastic hinges at D and F) yields the highest value of M_p and is therefore the true mechanism.

REDUCTION IN PLASTIC-MOMENT CAPACITY CAUSED BY AXIAL FORCE

A W10 × 45 beam-column is subjected to an axial force of 84 kips (373.6 kN) at ultimate load. (*a*) Applying the exact method, calculate the plastic moment this section can develop with respect to the major axis. (*b*) Construct the interaction diagram for this section, and then calculate the plastic moment by assuming a linear interaction relationship that approximates the true relationship.

Calculation Procedure:

1. Record the relevant properties of the member

Let P = applied axial force, kips (kN); P_y = axial force that would induce plastification if acting alone, kips (kN) = Af_y; M'_p = plastic-moment capacity of the section in combination with P, ft·kips (kN·m).

A typical stress diagram for a beam-column at plastification is shown in Fig. 93a. To simplify the calculations, resolve this diagram into the two parts shown at the right. This procedure is tantamount to assuming that the axial load is resisted by a central core and the moment by the outer segments of the section, although in reality they are jointly resisted by the integral action of the entire section.

From the AISC *Manual*, for a W10 × 45: A = 13.24 sq.in. (85.424 cm^2); d = 10.12 in. (257.048 mm); t_f = 0.618 in. (15.6972 mm); t_w = 0.350 in. (8.890 mm); d_w = 10.12 − 2(0.618) = 8.884 in. (225.6536 mm); Z = 55.0 in^3 (901.45 cm^3).

2. Assume that the central core that resists the 84-kip (373.6-kN) load is encompassed within the web; determine the core depth

Calling the depth of the core g, refer to Fig. 93d. Then g = 84/[0.35(36)] = 6.67 < 8.884 in. (225.6536 mm).

3. Compute the plastic modulus of the core, the plastic modulus of the remaining section, and the value of M'_p

Using data from the *Manual* for the plastic modulus of a rectangle, we find $Z_c = \frac{1}{4} t_w g^2$ = ¼(0.35)(6.67)^2 = 3.9 in^3 (63.92 cm^3); Z_r = 55.0 − 3.9 = 51.1 in^3 (837.53 cm^3); M'_p = 51.1(36)/12 = 153.3 ft·kips (207.87 kN·m). This constitutes the solution of part a. The solution of part b is given in steps 4 through 6.

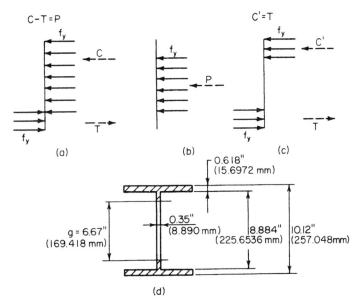

FIGURE 93

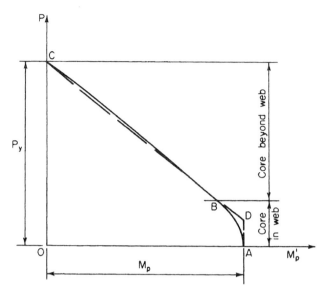

FIGURE 94. Interaction diagram for axial force and moment.

4. Assign a series of values to the parameter g, and compute the corresponding sets of values of P and M'_p

Apply the results to plot the interaction diagram in Fig. 94. This comprises the parabolic curves CB and BA, where the points A, B, and C correspond to the conditions $g = 0$, $g = d_w$, and $g = d$, respectively.

The interaction diagram is readily analyzed by applying the following relationships: $dP/dg = f_y t$; $dM'_p/dg = -\frac{1}{2}f_y tg$; $\therefore dP/dM'_p = -2/g$. This result discloses that the change in slope along CB is very small, and the curvature of this arc is negligible.

5. Replace the true interaction diagram with a linear one

Draw a vertical line $AD = 0.15P_y$, and then draw the straight line CD (Fig. 94). Establish the equation of CD. Thus, slope of $CD = -0.85P_y/M_p$; $P = -0.85P_y M'_p/M_p$, or $M'_p = 1.18(1 - P/P_y)M_p$.

The provisions of one section of the AISC *Specification* are based on the linear interaction diagram.

6. Ascertain whether the data are represented by a point on AD or CD; calculate M'_p, accordingly

Thus, $P_y = Af_y = 13.24(36) = 476.6$ kips (2119.92 kN); $P/P_y = 84/476.6 = 0.176$; therefore, apply the last equation given in step 5. Thus, $M_p = 55.0(36)/12 = 165$ ft·kips (223.7 kN·m); $M'_p = 1.18(1 - 0.176)(165) = 160.4$ ft·kips (217.50 kN·m). This result differs from that in part *a* by 4.6 percent.

PART 2

HANGERS, CONNECTORS, AND WIND-STRESS ANALYSIS

In the following Calculation Procedures, structural steel members are designed in accordance with the *Specification for the Design, Fabrication and Erection of Structural Steel for Buildings* of the American Institute of Steel Construction. In the absence of any statement to the contrary, it is to be understood that the structural-steel members are made of ASTM A36 steel, which has a yield-point stress of 36,000 lb/sq.in. (248.2 MPa).

Reinforced-concrete members are designed in accordance with the specification *Building Code Requirements for Reinforced Concrete* of the American Concrete Institute.

DESIGN OF AN EYEBAR

A hanger is to carry a load of 175 kips (778.4 kN). Design an eyebar of A440 steel.

Calculation Procedure:

1. Record the yield-point stresses of the steel

Refer to Fig. 1 for the notational system. Let subscripts 1 and 2 refer to cross sections through the body of the bar and through the center of the pin hole, respectively.

Eyebars are generally flame-cut from plates of high-strength steel. The design provisions of the AISC *Specification* reflect the results of extensive testing of such members. A section of the *Specification* permits a tensile stress of $0.60 f_y$ at 1 and $0.45 f_y$ at 2, where f_y denotes the yield-point stress.

From the AISC *Manual* for A440 steel:

If $t \leq 0.75$ in. (19.1 mm), $f_y = 50$ kips/sq.in. (344.7 MPa).
If $0.75 < t \leq 1.5$ in. (38 mm), $f_y = 46$ kips/sq.in. (317.1 MPa).
If $1.5 < t \leq 4$ in. (102 mm), $f_y = 42$ kips/sq.in. (289.5 MPa).

2. Design the body of the member, using a trial thickness

The *Specification* restricts the ratio w/t to a value of 8. Compute the capacity P of a ¾-in. (19.1-mm) eyebar of maximum width. Thus $w = 8(¾) = 6$ in. (152 mm); $f = 0.6(50) = 30$ kips/sq.in. (206.8 MPa); $P = 6(0.75)30 = 135$ kips (600.5 kN). This is not acceptable

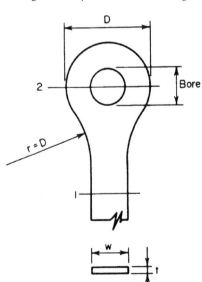

FIGURE 1. Eyebar hanger.

because the desired capacity is 175 kips (778.4 kN). Hence, the required thickness exceeds the trial value of ¾ in. (19.1 mm). With t greater than ¾ in. (19.1 mm), the allowable stress at 1 is $0.60f_y$, or 0.60(46 kips/sq.in.) = 27.6 kips/sq.in. (190.3 MPa); say 27.5 kips/sq.in. (189.6 MPa) for design use. At 2 the allowable stress is 0.45(46) = 20.7 kips/sq.in. (142.7 MPa), say 20.5 kips/sq.in. (141.3 MPa) for design purposes.

To determine the required area at 1, use the relation $A_1 = P/f$, where f = allowable stress as computed above. Thus, $A_1 = 175/27.5 = 6.36$ sq.in. (4103 mm²). Use a plate 6½ × 1 in. (165 × 25.4 mm) in which $A_1 = 6.5$ sq.in. (4192 mm²).

3. Design the section through the pin hole

The AISC *Specification* limits the pin diameter to a minimum value of $7w/8$. Select a pin diameter of 6 in. (152 mm). The bore will then be 6¹/₃₂ in. (153 mm) diameter. The net width required will be $P/(ft) = 175/[20.5(1.0)] = 8.54$ in. (217 mm); $D_{min} = 6.03 + 8.54 = 14.57$ in. (370 mm). Set $D = 14¾$ in. (375 mm), $A_2 = 1.0(14.75 - 6.03) = 8.72$ sq.in. (5626 mm²); $A_2/A_1 = 1.34$. This result is satisfactory, because the ratio of A_2/A_1 must lie between 1.33 and 1.50.

4. Determine the transition radius r

In accordance with the *Specification*, set $r = D = 14¾$ in. (374.7 mm).

ANALYSIS OF A STEEL HANGER

A 12 × ½ in. (305 × 12.7 mm) steel plate is to support a tensile load applied 2.2 in. (55.9 mm) from its center. Determine the ultimate load.

Calculation Procedure:

1. Determine the distance x

The plastic analysis of steel structures is developed in Sec. 1 of this handbook. Figure 2a is the load diagram, and Fig. 2b is the stress diagram at plastification. The latter may be replaced for convenience with the stress diagram in Fig. 2c, where $T_1 = C$; P_u = ultimate load; e = eccentricity; M_u = ultimate moment = $P_u e$; f_y = yield-point stress; d = depth of section; t = thickness of section.

By using Fig. 2c,

$$P_u = T_2 = f_y t(d - 2x) \tag{1}$$

Also, $T_1 = f_y tx$, and $M_u = P_u e = T_1(d - x)$, so

$$x = \frac{d}{2} + e - \left[\left(\frac{d}{2} + e\right)^2 - ed\right]^{0.5} \tag{2}$$

Or, $x = 6 + 2.2 - [(6 + 2.2)^2 - 2.2 \times 12]^{0.5} = 1.81$ in. (45.9 mm).

2. Find P_u

By Eq. 1, $P_u = 36,000(0.50)(12 - 3.62) = 151,000$ lb (671.6 kN).

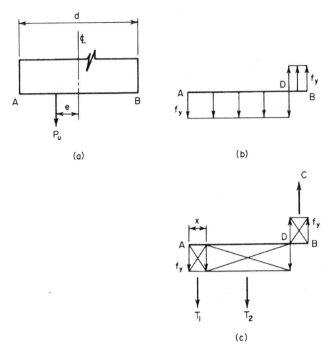

FIGURE 2

ANALYSIS OF A GUSSET PLATE

The gusset plate in Fig. 3 is ½ in. (12.7 mm) thick and connects three web members to the bottom chord of a truss. The plate is subjected to the indicated ultimate forces, and transfer

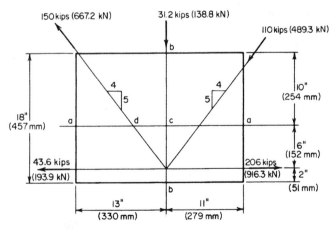

FIGURE 3. Gusset plate.

HANGERS, CONNECTORS, AND WIND-STRESS ANALYSIS

of these forces from the web members to the plate is completed at section *a-a*. Investigate the adequacy of this plate. Use 18,000 lb/sq.in. (124.1 MPa) as the yield-point stress in shear, and disregard interaction of direct stress and shearing stress in computing the ultimate-load and ultimate-moment capacity.

Calculation Procedure:

1. Resolve the diagonal forces into their horizontal and vertical components
Let H_u and V_u denote the ultimate shearing force on a horizontal and vertical plane, respectively. Resolving the diagonal forces into their horizontal and vertical components gives $(4^2 + 5^2)^{0.5} = 6.40$. Horizontal components: $150(4/6.40) = 93.7$ kips (416.8 kN); $110(4/6.40) = 68.7$ kips (305.6 kN). Vertical components: $150(5/6.40) = 117.1$ kips (520.9 kN); $110(5/6.40) = 85.9$ kips (382.1 kN).

2. Check the force system for equilibrium
Thus, $\Sigma F_H = 206.0 - 43.6 - 93.7 - 68.7 = 0$; this is satisfactory, as is $\Sigma F_V = 117.1 - 85.9 - 31.2 = 0$.

3. Compare the ultimate shear at section a-a with the allowable value
Thus, $H_u = 206.0 - 43.6 = 162.4$ kips (722.4 kN). To compute $H_{u,\text{allow}}$ assume that the shearing stress is equal to the yield-point stress across the entire section. Then $H_{u,\text{allow}} = 24(0.5)(18) = 216$ kips (960.8 kN). This is satisfactory.

4. Compare the ultimate shear at section b-b with the allowable value
Thus, $V_u = 117.1$ kips (520.9 kN); $V_{u,\text{allow}} = 18(0.5)(18) = 162$ kips (720.6 kN). This is satisfactory.

5. Compare the ultimate moment at section a-a with the plastic moment
Thus, $cd = 4(6)/5 = 4.8$ in. (122 mm); $M_u = 4.8(117.1 + 85.9) = 974$ in.·kips (110.1 kN·m). Or, $M_u = 6(206 - 43.6) = 974$ in.·kips (110.1 kN·m). To find the plastic moment M_p, use the relation $M_u = f_y bd^2/4$, or $M_p = 36(0.5)(24)^2/4 = 2592$ in.·kips (292.9 kN·m). This is satisfactory.

6. Compare the ultimate direct force at section b-b with the allowable value
Thus, $P_u = 93.7 + 43.6 = 137.3$ kips (610.7 kN); or $P_u = 206.0 - 68.7 = 137.3$ kips (610.7 kN); $e = 9 - 2 = 7$ in. (177.8 mm). By Eq. 2, $x = 9 + 7 - [(9 + 7)^2 - 7 \times 18]^{0.5} = 4.6$ in. (116.8 mm). By Eq. 1, $P_{u,\text{allow}} = 36,000(0.5)(18 - 9.2) = 158.4$ kips (704.6 kN). This is satisfactory.

On horizontal sections above *a-a*, the forces in the web members have not been completely transferred to the gusset plate, but the eccentricities are greater than those at *a-a*. Therefore, the calculations in step 5 should be repeated with reference to one or two sections above *a-a* before any conclusion concerning the adequacy of the plate is drawn.

DESIGN OF A SEMIRIGID CONNECTION

A W14 × 38 beam is to be connected to the flange of a column by a semirigid connection that transmits a shear of 25 kips (111.2 kN) and a moment of 315 in.·kips (35.6 kN·m). Design the connection for the moment, using A141 shop rivets and A325 field bolts of ⅞-in. (22.2-mm) diameter.

Calculation Procedure:

1. Record the relevant properties of the W14 × 38

A semirigid connection is one that offers only partial restraint against rotation. For a relatively small moment, a connection of the type shown in Fig. 4a will be adequate. In designing this type of connection, it is assumed for simplicity that the moment is resisted entirely by the flanges; and the force in each flange is found by dividing the moment by the beam depth.

Figure 4b indicates the assumed deformation of the upper angle, A being the point of contraflexure in the vertical leg. Since the true stress distribution cannot be readily ascertained, it is necessary to make simplifying assumptions. The following equations evolve from a conservative analysis of the member: $c = 0.6a$; $T_2 = T_1(1 + 3a/4b)$.

Study shows that use of an angle having two rows of bolts in the vertical leg would be unsatisfactory because the bolts in the outer row would remain inactive until those in the inner row yielded. If the two rows of bolts are required, the flange should be connected by means of a tee rather than an angle.

The following notational system will be used with reference to the beam dimensions: b = flange width; d = beam depth; t_f = flange thickness; t_w = web thickness.

Record the relevant properties of the W14 × 38; d = 14.12 in. (359 mm); t_f = 0.513 in. (13 mm). (Obtain these properties from a table of structural-shape data.)

2. Establish the capacity of the shop rivets and field bolts used in transmitting the moment

From the AISC *Specification*, the rivet capacity in single shear = 0.6013(15) = 9.02 kips (40.1 kN); rivet capacity in bearing 0.875(0.513)(48.5) = 21.77 kips (96.8 kN); bolt capacity in tension = 0.6013(40) = 24.05 kips (106.9 kN).

3. Determine the number of rivets required in each beam flange

Thus, T_1 = moment/d = 315/14.12 = 22.31 kips (99.7 kN); number of rivets = T_1/rivet capacity in single shear = 22.31/9.02 = 2.5; use four rivets, the next highest even number.

4. Assuming tentatively that one row of field bolts will suffice, design the flange angle

Try an angle 8 × 4 × ¾ in. (203 × 102 × 19 mm), 8 in. (203 mm) long, having a standard gage of 2½ in. (63.5 mm) in the vertical leg. Compute the maximum bending moment M in this leg. Thus, $c = 0.6(2.5 - 0.75) = 1.05$ in. (26.7 mm); $M = T_1 c = 23.43$ in.·kips (2.65 kN·m). Then apply the relation $f = M/S$ to find the flexural stress. Or, $f = 23.43/[(1/6)(8)(0.75)^2] = 31.24$ kips/sq.in. (215.4 MPa).

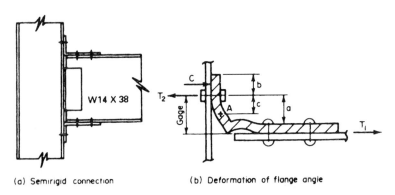

(a) Semirigid connection   (b) Deformation of flange angle

FIGURE 4

Since the cross section is rectangular, the allowable stress is 27 kips/sq.in. (186.1 MPa), as given by the AISC *Specification*. (The justification for allowing a higher flexural stress in a member of rectangular cross section as compared with a wide-flange member is presented in Sec. 1.)

Try a ⅞-in. (22-mm) angle, with $c = 0.975$ in. (24.8 mm); $M = 21.75$ in.·kips (2.46 kN·m); $f = 21.75/(1/6)(8)(0.875)^2 = 21.3$ kips/sq.in. (146.8 MPa). This is an acceptable stress.

5. Check the adequacy of the two field bolts in each angle
Thus, $T_2 = 22.31[1 + 3 \times 1.625/(4 \times 1.5)] = 40.44$ kips (179.9 kN); the capacity of two bolts $= 2(24.05) = 48.10$ kips (213.9 kN). Hence the bolts are acceptable because their capacity exceeds the load.

6. Summarize the design
Use angles $8 \times 4 \times ⅞$ in. ($203 \times 102 \times 19$ mm), 8 in. (203 mm) long. In each angle, use four rivets for the beam connection and two bolts for the column connection. For transmitting the shear, the standard web connection for a 14-in. (356-mm) beam shown in the AISC *Manual* is satisfactory.

RIVETED MOMENT CONNECTION

A W18 × 60 beam frames to the flange of a column and transmits a shear of 40 kips (177.9 kN) and a moment of 2500 in.·kips (282.5 kN·m). Design the connection, using ⅞-in. (22-mm) diameter rivets of A141 steel for both the shop and field connections.

Calculation Procedure:

1. Record the relevant properties of the W18 × 60
The connection is shown in Fig. 5a. Referring to the row of rivets in Fig. 5b, consider that there are n rivets having a uniform spacing p. The moment of inertia and section modulus of this rivet group with respect to its horizontal centroidal axis are

$$I = p^2 n \times \frac{n^2 - 1}{12} \qquad S = \frac{pn(n+1)}{6} \tag{3}$$

Record the properties of the W18 × 60: $d = 18.25$ in. (463.6 mm); $b = 7.558$ in. (192 mm); $k = 1.18$ in. (30.0 mm); $t_f = 0.695$ in. (17.7 mm); $t_w = 0.416$ in. (10.6 mm).

2. Establish the capacity of a rivet
Thus: single shear, 9.02 kips (40.1 kN); double shear, 18.04 kips (80.2 kN); bearing on beam web, $0.875(0.416)(48.5) = 17.65$ kips (78.5 kN).

3. Determine the number of rivets required on line 1 as governed by the rivet capacity
Try 15 rivets having the indicated disposition. Apply Eq. 3 with $n = 17$; then make the necessary correction. Thus, $I = 9(17)(17^2 - 1)/12 - 2(9)^2 = 3510$ sq.in. (22,645 cm²); $S = 3510/24 = 146.3$ in. (3716 mm).

Let F denote the force on a rivet, and let the subscripts x and y denote the horizontal and vertical components, respectively. Thus, $F_x = M/S = 2500/146.3 = 17.09$ kips (76.0 kN); $F_y = 40/15 = 2.67$ kips (11.9 kN); $F = (17.09^2 + 2.67^2)^{0.5} = 17.30 < 17.65$. Therefore, this is acceptable.

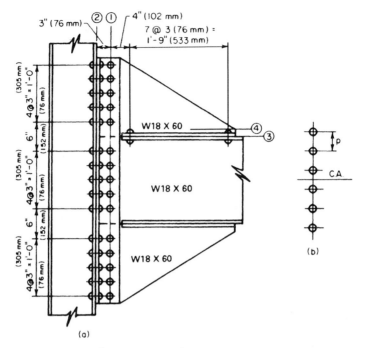

FIGURE 5. Riveted moment connection.

4. Compute the stresses in the web plate at line 1
The plate is considered continuous; the rivet holes are assumed to be 1 in. (25.4 mm) in diameter for the reasons explained earlier.

The total depth of the plate is 51 in. (1295.4 mm), the area and moment of inertia of the net section are $A_n = 0.416(51 - 15 \times 1) = 14.98$ sq.in. (96.6 cm^2) and $I_n = (1/12)(0.416)(51)^3 - 1.0(0.416)(3510) = 3138$ in^4 (130,603.6 cm^4).

Apply the general shear equation. Since the section is rectangular, the maximum shearing stress is $v = 1.5V/A_n = 1.5(40)/14.98 = 4.0$ kips/sq.in. (27.6 MPa). The AISC *Specification* gives an allowable stress of 14.5 kips/sq.in. (99.9 MPa).

The maximum flexural stress is $f = Mc/I_n = 2500(25.5)13138 = 20.3 < 27$ kips/sq.in. (186.1 MPa). This is acceptable. The use of 15 rivets is therefore satisfactory.

5. Compute the stresses in the rivets on line 2
The center of rotation of the angles cannot be readily located because it depends on the amount of initial tension to which the rivets are subjected. For a conservative approximation, assume that the center of rotation of the angles coincides with the horizontal centroidal axis of the rivet group. The forces are $F_x = 2500/[2(146.3)] = 8.54$ kips (37.9 kN); $F_y = 40/30 = 1.33$ kips (5.9 kN). The corresponding stresses in tension and shear are $s_t = F_y/A = 8.54/0.6013 = 14.20$ kips/sq.in. (97.9 MPa); $s_s = F_y/A = 1.33/0.6013 = 2.21$ kips/sq.in. (15.2 MPa). The *Specification* gives $s_{t,\text{allow}} = 28 - 1.6(2.21) > 20$ kips/sq.in. (137.9 kPa). This is acceptable.

6. Select the size of the connection angles
The angles are designed by assuming a uniform bending stress across a distance equal to the spacing p of the rivets; the maximum stress is found by applying the tensile force on the extreme rivet.

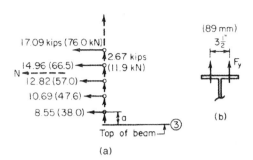

FIGURE 6

Try $4 \times 4 \times \frac{3}{4}$ in. ($102 \times 102 \times 19$ mm) angles, with a standard gage of $2\frac{1}{2}$ in. (63.5 mm) in the outstanding legs. Assuming the point of contraflexure to have the location specified in the previous calculation procedure, we get $c = 0.6(2.5 - 0.75) = 1.05$ in. (26.7 mm); $M = 8.54(1.05) = 8.97$ in.·kips (1.0 kN·m); $f = 8.97/[(^1/_6)(3)(0.75)^2] = 31.9 > 27$ kips/sq.in. (186.1 MPa). Use $5 \times 5 \times \frac{7}{8}$ in. ($127 \times 127 \times 22$ mm) angles, with a $2\frac{1}{2}$-in. (63.5-mm) gage in the outstanding legs.

7. Determine the number of rivets required on line 3

The forces in the rivets above this line are shown in Fig. 6a. The resultant forces are $H = 64.11$ kips (285.2 kN); $V = 13.35$ kips (59.4 kN). Let M_3 denote the moment of H with respect to line 3. Then $a = \frac{1}{2}(24 - 18.25) = 2.88$ in. (73.2 mm); $M_3 = 633.3$ in.·kips (71.6 kN·m).

With reference to Fig. 6b, the tensile force F_y in the rivet is usually limited by the bending capacity of beam flange. As shown in the AISC *Manual*, the standard gage in the W18 × 60 is $3\frac{1}{2}$ in. (88.9 mm). Assume that the point of contraflexure in the beam flange lies midway between the center of the rivet and the face of the web. Referring to Fig. 4b, we have $c = \frac{1}{2}(1.75 - 0.416/2) = 0.771$ in. (19.6 mm); $M_{allow} = fS = 27(^1/_6)(3)(0.695)^2 = 0.52$ in.·kips (0.74 kN·m). If the compressive force C is disregarded, $F_{y,allow} = 6.52/0.771 = 8.46$ kips (37.6 kN).

Try 16 rivets. The moment on the rivet group is $M = 633.3 - 13.35(14.5) = 440$ in.·kips (49.7 kN·m). By Eq. 3, $S = 2(3)(8)(9)/6 = 72$ in. (1829 mm). Also, $F_y = 440/72 + 13.35/16 = 6.94 < 8.46$ kips (37.6 kN). This is acceptable. (The value of F_y corresponding to 14 rivets is excessive.)

The rivet stresses are $s_t = 6.94/0.6013 = 11.54$ kips/sq.in. (79.6 MPa); $s_s = 64.11/[16(0.6013)] = 6.67$ kips/sq.in. (45.9 MPa). From the *Specification*, $s_{t,allow} = 28 - 1.6(6.67) = 17.33$ kips/sq.in. (119.5 MPa). This is acceptable. The use of 16 rivets is therefore satisfactory.

8. Compute the stresses in the bracket at the toe of the fillet (line 4)

Since these stresses are seldom critical, take the length of the bracket as 24 in. (609.6 mm) and disregard the eccentricity of V. Then $M = 633.3 - 64.11(1.18) = 558$ in.·kips (63.1 kN·m); $f = 558/[(^1/_6)(0.416)(24)^2] + 13.35/[0.416(24)] = 15.31$ kips/sq.in. (105.5 MPa). This is acceptable. Also, $v = 1.5(64.11)/[0.416(24)] = 9.63$ kips/sq.in. (66.4 MPa) This is also acceptable.

DESIGN OF A WELDED FLEXIBLE BEAM CONNECTION

A W18 × 64 beam is to be connected to the flange of its supporting column by means of a welded framed connection, using E60 electrodes. Design a connection to transmit a reaction of 40 kips (177.9 kN). The AISC table of welded connections may be applied in selecting the connection, but the design must be verified by computing the stresses.

Calculation Procedure:

1. Record the pertinent properties of the beam
It is necessary to investigate both the stresses in the weld and the shearing stress in the beam induced by the connection. The framing angles must fit between the fillets of the beam. Record the properties: $T = 15\frac{3}{8}$ in. (390.5 mm); $t_w = 0.403$ in. (10.2 mm).

2. Select the most economical connection from the AISC Manual
The most economical connection is: angles $3 \times 3 \times 5/16$ in. ($76 \times 76 \times 7.9$ mm), 12 in. (305 mm) long; weld size $> 3/16$ in. (4.8 mm) for connection to beam web, $\frac{1}{4}$ in. (6.4 mm) for connection to the supporting member.

According to the AISC table, weld A has a capacity of 40.3 kips (179.3 kN), and weld B has a capacity of 42.8 kips (190.4 kN). The minimum web thickness required is 0.25 in. (6.4 mm). The connection is shown in Fig. 7a.

3. Compute the unit force in the shop weld
The shop weld connects the angles to the beam web. Refer to Sec. 1 for two calculation procedures for analyzing welded connections.

The weld for one angle is shown in Fig. 7b. The allowable force, as given in Sec. 1, is $m = 2(2.5)(1.25)/[2(2.5)+12] = 0.37$ in. (9.4 mm); $P = 20,000$ lb (88.9 kN); $M = 20,000(3-0.37) = 52,600$ in.·lb (5942.7 N·m); $I_x = (1/12)(12)^3 + 2(2.5)(6)^2 = 324$ in^3 (5309.4 cm^3); $I_y = 12(0.37)^2 + 2(1/12)(2.5)^3 + 2(2.5)(0.88)^2 = 8$ in^3 (131.1 cm^3); $J = 324 + 8 = 332$ in^3 (5440.5 cm^3); $f_x = My/J = 52,600(6)/332 = 951$ lb/lin in. (166.5 N/mm); $f_y = Mx/J = 52,600(2.5)(0.37)/332 = 337$ lb/lin in. (59.0 N/mm); $f_y = 20,000/(2 \times 2.5 + 12) = 1176$ lb/lin in. (205.9 N/mm); $F_x = 951$ lb/lin in. (166.5 N/mm); $F_y = 337 + 1176 = 1513$ lb/lin in. (265.0 N/mm); $F = (951^2 + 1513^2)^{0.5} = 1787 < 1800$, which is acceptable.

4. Compute the shearing stress in the web
The allowable stress given in the AISC *Manual* is 14,500 lb/sq.in. (99.9 MPa). The two angles transmit a unit shearing force of 3574 lb/lin in. (0.64 kN/mm) to the web. The shearing stress is $v = 3574/0.403 = 8870$ lb/sq.in. (61.1 MPa), which is acceptable.

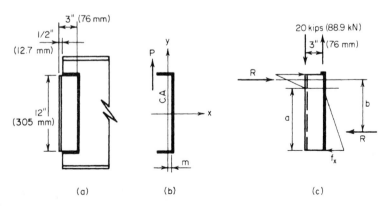

FIGURE 7. Welded flexible beam connection.

5. Compute the unit force in the field weld

The field weld connects the angles to the supporting member. As a result of the 3-in. (76.2-mm) eccentricity on the outstanding legs, the angles tend to rotate about a neutral axis located near the top, bearing against the beam web above this axis and pulling away from the web below this axis. Assume that the distance from the top of the angle to the neutral axis is one-sixth of the length of the angle. The resultant forces are shown in Fig. 7c. Then $a = (5/6)12 = 10$ in. (254 mm); $b = (2/3)12 = 8$ in. (203 mm); $B = 20,000(3)/8 = 7500$ lb (33.4 kN); $f_x = 2R/a = 1500$ lb/lin in. (262.7 N/mm); $f_y = 20,000/12 = 1667$ lb/lin in. (291.9 N/mm); $F (1500^2 + 1667^2)^{0.5} = 2240 < 2400$ lb/lin in. (420.3 N/mm), which is acceptable. The weld is returned a distance of ½ in. (12.7 mm) across the top of the angle, as shown in the AISC *Manual*.

DESIGN OF A WELDED SEATED BEAM CONNECTION

A W27 × 94 beam with a reaction of 77 kips (342.5 kN) is to be supported on a seat. Design a welded connection, using E60 electrodes.

Calculation Procedure:

1. Record the relevant properties of the beam

Refer to the AISC *Manual*. The connection will consist of a horizontal seat plate and a stiffener plate below the seat, as shown in Fig. 8a. Record the relevant properties of the W27 × 94: $k = 1.44$ in. (36.6 mm); $b = 9.99$ in. (253.7 mm); $t_f = 0.747$ in. (19.0 mm); $t_w = 0.490$ in. (12.4 mm).

2. Compute the effective length of bearing

Equate the compressive stress at the toe of the fillet to its allowable value of 27 kips/sq.in. (186.1 MPa) as given in the AISC *Manual*. Assume that the reaction distributes itself through the web at an angle of 45°. Refer to Fig. 8b. Then $N = P/27t_w - k$, or $N = 77/27(0.490) - 1.44 = 4.38$ in. (111.3 mm).

3. Design the seat plate

As shown in the AISC *Manual*, the beam is set back about ½ in. (12.7 mm) from the face of the support. Make $W = 5$ in. (127.0 mm). The minimum allowable distance from the edge of the seat plate to the edge of the flange equals the weld size plus 5/16 in. (7.8 mm). Make the seat plate 12 in. (304.8 mm) long; its thickness will be made the same as that of the stiffener.

4. Design the weld connecting the stiffener plate to the support

The stresses in this weld are not amenable to precise analysis. The stiffener rotates about a neutral axis, bearing against the support below this axis and pulling away from the support above this axis. Assume for simplicity that the neutral axis coincides with the centroidal axis of the weld group; the maximum weld stress occurs at the top. A weld length of 0.2L is supplied under the seat plate on each side of the stiffener. Refer to Fig. 8c.

Compute the distance e from the face of the support to the center of the bearing, measuring N from the edge of the seat. Thus, $e = W - N/2 = 5 - 4.38/2 = 2.81$ in. (71.4 mm); $P = 77$ kips (342.5 kN); $M = 77(2.81) = 216.4$ in.·kips (24.5 kN·m); $m = 0.417L$; $I_x = 0.25L^3 f_1 = Mc/I_x = 216.4(0.417L)/0.25L^3 = 361.0/L^2$ kips/lin in.; $f_2 = P/A = 77/2.4L = 32.08/L$ kips/lin in. Use a 5/16-in. (7.9-mm) weld, which has a capacity of 3 kips/lin in. (525.4 N/mm). Then $F^2 = f_1^2 + f_2^2 = 130,300/L^4 + 1029/L^2 \le 3^2$. This equation is satisfied by $L = 14$ in. (355.6 mm).

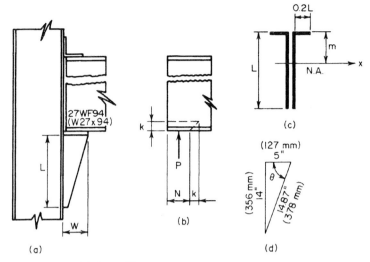

FIGURE 8. Welded seated beam connection.

5. Determine the thickness of the stiffener plate

Assume this plate is triangular (Fig. 8*d*). The critical section for bending is assumed to coincide with the throat of the plate, and the maximum bending stress may be obtained by applying $f = (P/tW \sin^2 \theta)(1 + 6e'/W)$, where e' = distance from center of seat to center of bearing.

Using an allowable stress of 22,000 lb/sq.in. (151.7 MPa), we have $e' = e - 2.5 = 0.31$ in. (7.9 mm), $t = \{77/[22 \times 5(14/14.87)^2]\}(1 + 6 \times 0.31/5) = 1.08$ in. (27.4 mm).

Use a 1⅛-in. (28.6-mm) stiffener plate. The shearing stress in the plate caused by the weld is $v = 2(3000)/1.125 = 5330 < 14,500$ lb/sq.in. (99.9 MPa), which is acceptable.

DESIGN OF A WELDED MOMENT CONNECTION

A W16 × 40 beam frames to the flange of a W12 × 72 column and transmits a shear of 42 kips (186.8 kN) and a moment of 1520 in.·kips (171.1 kN·m). Design a welded connection, using E60 electrodes.

Calculation Procedure:

1. Record the relevant properties of the two sections

In designing a welded moment connection, it is assumed for simplicity that the beam flanges alone resist the bending moment. Consequently, the beam transmits three forces to the column: the tensile force in the top flange, the compressive force in the bottom flange, and the vertical load. Although the connection is designed ostensibly on an elastic design basis, it is necessary to consider its behavior at ultimate load, since a plastic hinge would form at this joint. The connection is shown in Fig. 9.

Record the relevant properties of the sections: for the W16 × 40, $d = 16.00$ in. (406.4 mm); $b = 7.00$ in. (177.8 mm); $t_f = 0.503$ in. (12.8 mm); $t_w = 0.307$ in. (7.8 mm); $A_f = 7.00(0.503) = 3.52$ sq.in. (22.7 cm²). For the W12 × 72, $k = 1.25$ in. (31.8 mm); $t_f = 0.671$ in. (17.04 mm); $t_w = 0.403$ in. (10.2 mm).

2. Investigate the need for column stiffeners: design the stiffeners if they are needed

The forces in the beam flanges introduce two potential modes of failure: crippling of the column web caused by the compressive force, and fracture of the weld transmitting the tensile force as a result of the bending of the column flange. The AISC *Specification* establishes the criteria for ascertaining whether column stiffeners are required. The first criterion is obtained by equating the compressive stress in the column web at the toe of the fillet to the yield-point stress f_y; the second criterion was obtained empirically. At the ultimate load, the capacity of the unreinforced web = $(0.503 + 5 \times 1.25)0.430 f_y = 2.904 f_y$; capacity of beam flange = $3.52 f_y$; $0.4(A_f)^{0.5} = 0.4(3.52)^{0.5} = 0.750 > 0.671$ in. (17.04 mm).

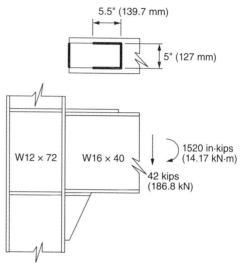

FIGURE 9. Welded moment connection.

Stiffeners are therefore required opposite both flanges of the beam. The required area is $A_{st} = 3.52 − 2.904 = 0.616$ sq.in. (3.97 cm²). Make the stiffener plates 3½ in. (88.9 mm) wide to match the beam flange. From the AISC, $t_{min} = 3.5/8.5 = 0.41$ in. (10.4 mm). Use two 3½ × ½ in. (88.9 × 12.7 mm) stiffener plates opposite both beam flanges.

3. Design the connection plate for the top flange

Compute the flange force by applying the total depth of the beam. Thus, $F = 1520/16.00 = 95$ kips (422.6 kN); $A = 95/22 = 4.32$ sq.in. (27.87 cm²).

Since the beam flange is 7 in. (177.8 mm) wide, use a plate 5 in. (127 mm) wide and ⅞ in. (22.2 mm) thick, for which $A = 4.38$ sq.in. (28.26 cm²). This plate is butt-welded to the column flange and fillet-welded to the beam flange. In accordance with the AISC *Specification*, the minimum weld size is 5/16 in. (7.94 mm) and the maximum size is 13/16 in. (20.6 mm). Use a ⅝-in. (15.9-mm) weld, which has a capacity of 6000 lb/lin in. (1051 N/mm). Then, length of weld = 95/6 = 15.8 in. (401.3 mm), say 16 in. (406.4 mm). To ensure that yielding of the joint at ultimate load will occur in the plate rather than in the weld, the top plate is left unwelded for a distance approximately equal to its width, as shown in Fig. 9.

4. Design the seat

The connection plate for the bottom flange requires the same area and length of weld as does the plate for the top flange. The stiffener plate and its connecting weld are designed in the same manner as in the previous calculation procedure.

RECTANGULAR KNEE OF RIGID BENT

Figure 10a is the elevation of the knee of a rigid bent. Design the knee to transmit an ultimate moment of 8100 in.·kips (914.5 kN·m).

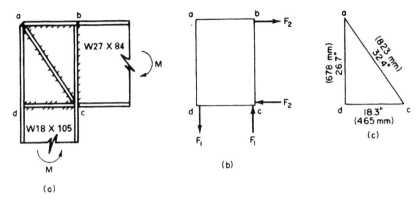

FIGURE 10. Rectangular knee.

Calculation Procedure:

1. Record the relevant properties of the two sections

Refer to the AISC *Specification* and *Manual*. It is assumed that the moment in each member is resisted entirely by the flanges and that the distance between the resultant flange forces is 0.95 times the depth of the member.

Record the properties of the members: for the W18 × 105, $d = 18.32$ in. (465.3 mm); $b_f = 11.79$ in. (299.5 mm); $t_f = 0.911$ in. (23.1 mm); $t_w = 0.554$ in. (14.1 mm); $k = 1.625$ in. (41.3 mm). For the W27 × 84, $d = 26.69$ in. (677.9 mm); $b_f = 9.96$ in. (253 mm); $t_f = 0.636$ in. (16.2 mm); $t_w = 0.463$ in. (11.8 mm).

2. Compute F_1

Thus, $F_1 M_u/(0.95d) = 8100/[0.95(18.32)] = 465$ kips (2068.3 kN).

3. Determine whether web stiffeners are needed to transmit F_1

The shearing stress is assumed to vary linearly from zero at a to its maximum value at d. The allowable average shearing stress is taken as $f_y/(3)^{0.5}$, where f_y denotes the yield-point stress. The capacity of the web = $0.554(26.69)(36/3^{0.5}) = 307$ kips (1365.5 kN). Therefore, use diagonal web stiffeners.

4. Design the web stiffeners

Referring to Fig. 10c, we see that $ac = (18.3^2 + 26.7^2)^{0.5} = 32.4$ in. (823 mm). The force in the stiffeners = $(465 - 307)32.4/26.7 = 192$ kips (854.0 kN). (The same result is obtained by computing F_2 and considering the capacity of the web across ab.) Then, $A_{st} = 192/36 = 5.33$ sq.in. (34.39 cm²). Use two plates 4 × ¾ in. (101.6 × 19.1 mm).

5. Design the welds, using E60 electrodes

The AISC *Specification* stipulates that the weld capacity at ultimate load is 1.67 times the capacity at the working load. Consequently, the ultimate-load capacity is 1000 lb/lin in. (175 N/mm) times the number of sixteenths in the weld size. The welds are generally designed to develop the full moment capacity of each member. Refer to the AISC *Specification*.

Weld at ab. This weld transmits the force in the flange of the 27-in. (685.8-mm) member to the web of the 18-in. (457.2-mm) member. Then $F = 9.96(0.636)(36) = 228$ kips (1014.1 kN), weld force = $228/[2(d - 2t_f)] = 228/[2(18.32 - 1.82)] = 6.91$ kips/lin in. (1210.1 N/mm). Use a ⁷⁄₁₆-in. (11.1-mm) weld.

Weld at bc. Use a full-penetration butt weld.

Weld at ac. Use the minimum size of ¼ in. (6.4 mm). The required total length of weld is $L = 192/4 = 48$ in. (1219.2 mm).

Weld at dc. Let F_3 denote that part of F_2 that is transmitted to the web of the 18-in. (457.2-mm) member through bearing, and let F_4 denote the remainder of F_2. Force F_3 distributes itself through the 18-in. (457.2-mm) member at 45° angles, and the maximum compressive stress occurs at the toe of the fillet. Find F_3 by equating this stress to 36 kips/sq.in. (248.2 MPa); or $F_3 = 36(0.554)(0.636 + 2 \times 1.625) = 78$ kips (346.9 kN). To evaluate F_4, apply the moment capacity of the 27-in. (685.8-mm) member. Or $F_4 = 228 - 78 = 150$ kips (667.2 kN).

The minimum weld size of ¼ in. (6.4 mm) is inadequate. Use a ⁵/₁₆-in. (7.9-mm) weld. The required total length is $L = 150/5 = 30$ in. (762.0 mm).

CURVED KNEE OF RIGID BENT

In Fig. 11 the rafter and column are both W21 × 82, and the ultimate moment at the two sections of tangency—*p* and *q*—is 6600 in.·kips (745.7 kN·m). The section of contraflexure in each member lies 84 in. (2133.6 mm) from the section of tangency. Design the knee.

Calculation Procedure:

1. Record the relevant properties of the members

Refer to the Commentary in the AISC *Manual.* The notational system is the same as that used in the *Manual,* plus a = distance from section of contraflexure to section of tangency; b = member flange width; x = distance from section of tangency to given section; M = ultimate moment at given section; M_p = plastic-moment capacity of knee at the given section.

Assume that the moment gradient dM/dx remains constant across the knee. The web thickness of the knee is made equal to that of the main material. The flange thickness of the knee, however, must exceed that of the main material, for this reason: As x increases, both M and M_p increase, but the former increases at a faster rate when x is small. The critical section occurs where $dM/dx = dM_p/dx$.

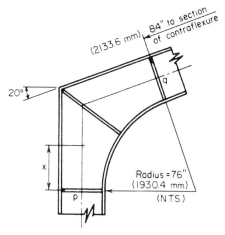

FIGURE 11. Curved knee.

An exact solution to this problem is possible, but the resulting equation is rather cumbersome. An approximate solution is given in the AISC *Manual.*

Record the relevant properties of the the W21 × 82: $d = 20.86$ in. (529.8 mm); $b = 8.96$ in. (227.6 mm); $t_f = 0.795$ in. (20.2 mm); $t_w = 0.499$ in. (12.7 mm).

2. Design the cross section of the knee, assuming tentatively that flexure is the sole criterion

Use a trial thickness of ½ in. (12.7 mm) for the web plate and a 9-in. (228.6-mm) width for the flange plate. Then $a = 84$ in. (2133.6 mm); $n = a/d = 84/20.86 = 4.03$. From the AISC *Manual,* $m = 0.14 \pm t' = t(1 + m) = 0.795(1.14) = 0.906$ in. (23.0 mm). Make the flange plate 1 in. (25.4 mm) thick.

3. Design the stiffeners; investigate the knee for compliance with the AISC Commentary

From the Commentary, *item 5*: Provide stiffener plates at the sections of tangency and at the center of the knee. Make the stiffener plates 4 × ⅞ in. (102 × 22 mm), one on each side of the web.

Item 3: Thus, $\phi = ½(90° − 20°) = 35°$; $\phi = 35/57.3 = 0.611$ rad; $L = R\phi = 76(0.611) = 46.4$ in. (1178.6 mm); or $L = \pi R(70°/360°) = 46.4$ in. (1178.6 mm); $L_{cr} = 6b = 6(9) = 54$ in. (1373 mm), which is acceptable.

Item 4: Thus, $b/t' = 9$; $2R/b = 152/9 = 16.9$, which is acceptable.

BASE PLATE FOR STEEL COLUMN CARRYING AXIAL LOAD

A W14 × 53 column carries a load of 240 kips (1067.5 kN) and is supported by a footing made of 3000-lb/sq.in. (20,682-kPa) concrete. Design the column base plate.

Calculation Procedure:

1. Compute the required area of the base plate; establish the plate dimensions

Refer to the base-plate diagram in the AISC *Manual*. The column load is assumed to be uniformly distributed within the indicated rectangle, and the footing reaction is assumed to be uniformly distributed across the base plate. The required thickness of the plate is established by computing the bending moment at the circumference of the indicated rectangle. Let f = maximum bending stress in plate; p = bearing stress; t = thickness of plate.

The ACI *Code* permits a bearing stress of 750 lb/sq.in. (5170.5 kPa) if the entire concrete area is loaded and 1125 lb/sq.in. (7755.8 kPa) if one-third of this area is loaded. Applying the 750-lb/sq.in. (5170.5-kPa) value, we get plate area = load, lb/750 = 240,000/750 = 320 sq.in. (2064.5 cm²).

The dimensions of the W14 × 53 are $d = 13.94$ in. (354.3 mm); $b = 8.06$ in. (204.7 mm); $0.95d = 13.24$ in. (335.3 mm); $0.80b = 6.45$ in. (163.8 mm). For economy, the projections m and n should be approximately equal. Set $B = 15$ in. (381 mm) and $C = 22$ in. (558.8 mm); then, area = 15(22) = 330 sq.in. (2129 cm²); $p = 240,000/330 = 727$ lb/sq.in. (5011.9 kPa).

2. Compute the required thickness of the base plate

Thus, $m = ½(22 − 13.24) = 4.38$ in. (111.3 mm), which governs. Also, $n = ½(15 − 6.45) = 4.28$ in. (108.7 mm).

The AISC *Specification* permits a bending stress of 27,000 lb/sq.in. (186.1 MPa) in a rectangular plate. The maximum bending stress is $f = M/S = 3pm^2/t^2$; $t = m(3p/f)^{0.5} = 4.38(3 \times 727/27,000)^{0.5} = 1.24$ in. (31.5 mm).

3. Summarize the design

Thus, $B = 15$ in. (381 mm); $C = 22$ in. (558.8 mm); $t = 1¼$ in. (31.8 mm).

BASE FOR STEEL COLUMN WITH END MOMENT

A steel column of 14-in. (355.6-mm.) depth transmits to its footing an axial load of 30 kips (133.4 kN) and a moment of 1100 in.·kips (124.3 kN·m) in the plane of its web. Design the base, using A307 anchor bolts and 3000-lb/sq.in. (20.7-MPa) concrete.

Calculation Procedure:

1. Record the allowable stresses and modular ratio
Refer to Fig. 12. If the moment is sufficiently large, it causes uplift at one end of the plate and thereby induces tension in the anchor bolt at that end. A rigorous analysis of the stresses in a column base transmitting a moment is not possible. For simplicity, compute the stresses across a horizontal plane through the base plate by treating this as the cross section of a reinforced-concrete beam, the anchor bolt on the tension side acting as the reinforcing steel. The effects of initial tension in the bolts are disregarded.

The anchor bolts are usually placed 2½ (63.5 mm) or 3 in. (76.2 mm) from the column flange. Using a plate of 26-in. (660-mm) depth as shown in Fig. 12a, let A_s = anchor-bolt cross-sectional area; B = base-plate width; C = resultant compressive force on base plate; T = tensile force in anchor bolt; f_s = stress in anchor bolt; p = maximum bearing stress; p' = bearing stress at column face; t = base-plate thickness.

Recording the allowable stresses and modular ratio by using the ACI *Code*, we get p = 750 lb/sq.in. (5170 kPa) and n = 9. From the AISC *Specification*, f_s = 14,000 lb/sq.in. (96.5 MPa); the allowable bending stress in the plate is 27,000 lb/sq.in. (186.1 MPa).

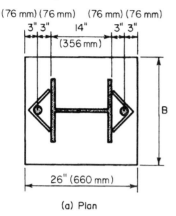
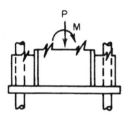
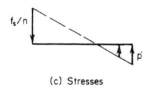

FIGURE 12. Anchor-bolt details. (a) Plan; (b) elevation; (c) stresses.

2. Construct the stress and force diagrams
These are shown in Fig. 13. Then f_s/n = 14/9 = 1.555 kips/sq.in. (10.7 MPa); kd = 23(0.750/2.305) = 7.48 in. (190.0 mm); jd = 23 − 7.48/3 = 20.51 in. (521.0 mm).

3. Design the base plate
Thus, C = ½(7.48)(0.750B) = 2.805B. Take moments with respect to the anchor bolt, or ΣM = 30(10) + 1100 − 2.805B(20.51) = 0; B = 24.3 in. (617.2 mm).

Assume that the critical bending stress in the base plate occurs at the face of the column. Compute the bending moment at the face for a 1-in. (25.4-mm) width of plate. Referring to Fig. 13c, we have p' = 0.750(1.48/7.48) = 0.148 kips/sq.in. (1020.3 kPa); M = $(6^2/6)$(0.148 + 2 × 0.750) = 9.89 in.·kips (1.12 kN·m); t^2 = 6M/27 = 2.20 sq.in. (14.19 cm²); t = 1.48 in. (37.6 mm). Make the base plate 25 in. (635 mm) wide and 1½ in. (38.1 mm) thick.

4. Design the anchor bolts
From the calculation in step 3, C = 2.805B = 2.805(24.3) = 68.2 kips (303.4 kN); T = 68.2 − 30 = 38.2 kips (169.9 kN); A_s = 38.2/14 = 2.73 sq.in. (17.61 cm²). Refer to the AISC *Manual*. Use 2¼-in. (57.2-mm) anchor bolts, one on each side of the flange. Then A_s = 3.02 sq.in. (19.48 cm²).

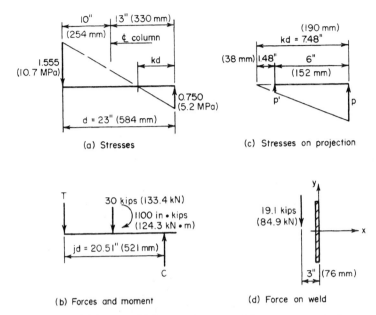

FIGURE 13. (*a*) Stresses; (*b*) forces and moment; (*c*) stresses on projection; (*d*) force on weld.

5. Design the anchorage for the bolts

The bolts are held by angles welded to the column flange, as shown in Fig. 12 and in the AISC *Manual*. Use ½-in. (12.7-mm) angles 12 in. (304.8 mm) long. Each line of weld resists a force of ½T. Refer to Fig. 13d and compute the unit force F at the extremity of the weld. Thus, $M = 19.1(3) = 57.3$ in.·kips (6.47 kN·m); $S_x = (1/6)(12)^2$ 24 sq.in. (154.8 cm²); $F_x = 57.3/24 = 2.39$ kips/lin in. (0.43 kN/mm); $F_y = 19.1/12 = 1.59$ kips/lin in (0.29 kN/mm); $F = (2.39^2 + 1.59^2)^{0.5} = 2.87$ kips/lin in. (0.52 kN/mm). Use a ⁵⁄₁₆-in. (4.8-mm) fillet weld of E60 electrodes, which has a capacity of 3 kips/lin in. (0.54 kN/mm).

GRILLAGE SUPPORT FOR COLUMN

A steel column in the form of a W14 × 320 reinforced with two 20 × 1½ in. (508 × 38.1 mm) cover plates carries a load of 2790 kips (12,410 kN). Design the grillage under this column, using an allowable bearing stress of 750 lb/sq.in. (5170.5 kPa) on the concrete. The space between the beams will be filled with concrete.

Calculation Procedure:

1. Establish the dimensions of the grillage

Refer to Fig. 14. A load of this magnitude cannot be transmitted from the column to its footing through the medium of a base plate alone. It is therefore necessary to interpose steel beams between the base plate and the footing; these may be arranged in one tier or in two orthogonal tiers. Integrity of each tier is achieved by tying the beams together by pipe separators. This type of column support is termed a *grillage*. In designing the grillage, it is assumed that bearing pressures are uniform across each surface under consideration.

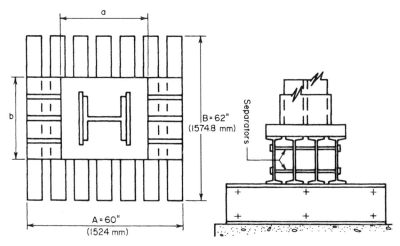

FIGURE 14. Grillage under column.

The area of grillage required = load, kips/allowable stress, kips/sq.in. = 2790/0.750 = 3720 sq.in. (23,994 cm²). Set $A = 60$ in. (1524 mm) and $B = 62$ in. (1574.8 mm), giving an area of 3720 sq.in. (23,994 cm²), as required.

2. Design the upper-tier beams

There are three criteria: bending stress, shearing stress, and compressive stress in the web at the toe of the fillet. The concrete between the beams supplies lateral restraint, and the allowable bending stress is therefore 24 kips/sq.in. (165.5 MPa).

Since the web stresses are important criteria, a grillage is generally constructed of S shapes rather than wide-flange beams to take advantage of the thick webs of S shapes. The design of the beams requires the concurrent determination of the length a of the base plate. Let f = bending stress; f_b, = compressive stress in web at fillet toe; v = shearing stress; P = load carried by single beam; S = section modulus of single beam; k = distance from outer surface of beam to toe of fillet; t_w = web thickness of beam; a_1 = length of plate as governed by flexure; a_2 = length of plate as governed by compressive stress in web.

Select a beam size on the basis of stresses f and f_b, and then investigate v. The maximum bending moment occurs at the center of the span; its value is $M = P(A - a)/8 = fS$; therefore, $a_1 = A - 8fS/P$.

At the toe of the fillet, the load P is distributed across a distance $a + 2k$. Then $f_b = P/(a + 2k)t_w$; therefore, $a_2 = P/f_b t_w - 2k$. Try four beams; then $P = 2790/4 = 697.5$ kips (3102.5 kN); $f = 24$ kips/sq.in. (165.5 MPa); $f_b = 27$ kips/sq.in. (186.1 MPa). Upon substitution, the foregoing equations reduce to $a_1 = 60 - 0.2755$; $a_2 = 25.8/t_w - 2k$.

Select the trial beam sizes shown in the accompanying table, and calculate the corresponding values of a_1 and a_2.

Size	S, in³ (cm³)	t_w, in. (mm)	k, in. (mm)	a_1, in. (mm)	a_2, in. (mm)
S18 × 54.7	88.4 (1448.6)	0.460 (11.68)	1.375 (34.93)	35.7 (906.8)	53.3 (1353.8)
S18 × 70	101.9 (1669.8)	0.711 (18.06)	1.375 (34.93)	32.0 (812.8)	33.6 (853.4)
S20 × 65.4	116.9 (1915.7)	0.500 (12.7)	1.563 (39.70)	27.9 (708.7)	48.5 (1231.9)
S20 × 75	126.3 (2069.7)	0.641 (16.28)	1.563 (39.70)	25.3 (642.6)	37.1 (942.9)

Try S18 × 70, with $a = 34$ in. (863.6 mm). The flange width is 6.25 in. (158.8 mm). The maximum vertical shear occurs at the edge of the plate; its magnitude is $V = P(A - a)(2A) = 697.5(60 - 34)/[2(60)] = 151.1$ kips (672.1 kN); $v = 151.1/[18(0.711)] = 11.8 < 14.5$ kips/sq.in. (99.9 MPa), which is acceptable.

3. Design the base plate
Refer to the second previous calculation procedure. To permit the deposition of concrete, allow a minimum space of 2 in. (50.8 mm) between the beam flanges. The minimum value of b is therefore $b = 4(6.25) + 3(2) = 31$ in. (787.4 mm).

The dimensions of the effective bearing area under the column are $0.95(16.81 + 2 \times 15) = 18.82$ in. (478.0 mm); $0.80(20) = 16$ in. (406.4 mm). The projections of the plate are $(34 - 18.82)/2 = 7.59$ in. (192.8 mm); $(31 - 16)/2 = 7.5$ in. (190.5 mm).

Therefore, keep $b = 31$ in. (787 mm), because this results in a well-proportioned plate. The pressure under the plate $= 2790/[34(31)] = 2.65$ kips/sq.in. (18.3 MPa). For a 1-in. (25.4-mm) width of plate, $M = \frac{1}{2}(2.65)/(7.59)^2 = 76.33$ in.·kips (8.6 kN·m); $S = M/f = 76.33/27 = 2.827$ in.^{3} (46.33 cm^3); $t = (6S)^{0.5} = 4.12$ in. (104.6 mm).

Plate thicknesses within this range vary by ⅛-in. (3.2-mm) increments, as stated in the AISC *Manual*. However, a section of the AISC *Specification* requires that plates over 4 in. (102 mm) thick be planed at all bearing surfaces. Set $t = 4½$ in. (114.3 mm) to allow for the planing.

4. Design the beams at the lower tier
Try seven beams. Thus, $P = 2790/7 = 398.6$ kips (1772.9 kN); $M = 398.6(62 - 31)/8 = 1545$ in.·kips (174.6 kN·m); $S_3 = 1545/24 = 64.4$ in^3 (1055.3 cm^3).

Try S15 × 50. Then $S = 64.2$ in^3 (1052.1 cm^3); $t_w = 0.550$ in. (14.0 mm); $k = 1.25$ in. (31.8 mm); $b = 5.64$ in. (143.3 mm). The space between flanges is $[60 - 7 \times 5.641]/6 = 3.42$ in. (86.9 mm). This result is satisfactory. Then $f_b = 398.6/[0.550(31 + 2 \times 1.25)] = 21.6 < 27$ kips/sq.in. (186.1 MPa), which is satisfactory; $V = 398.6(62 - 31)/[2(62)] = 99.7$ kips (443.5 kN); $v = 99.7/[15(0.550)] = 12.1 < 14.5$, which is satisfactory.

5. Summarize the design
Thus: $A = 60$ in. (1524 mm); $B = 62$ in. (1574.8 mm); base plate is $31 \times 34 \times 4½$ in. ($787.4 \times 863.6 \times 114.3$ mm), upper-tier steel, four beams S18 × 70; lower-tier steel, seven beams 15150.0.

WIND-STRESS ANALYSIS BY PORTAL METHOD

The bent in Fig. 15 resists the indicated wind loads. Applying the portal method of analysis, calculate all shears, end moments, and axial forces.

Calculation Procedure:

1. Compute the shear factor for each column
The portal method is an approximate and relatively simple method of wind-stress analysis that is frequently applied to regular bents of moderate height. It considers the bent to be composed of a group of individual portals and makes the following assumptions. (1) The wind load is distributed among the aisles of the bent in direct proportion to their relative widths. (2) The point of contraflexure in each member lies at its center.

Because of the first assumption, the shear in a given column is directly proportional to the average width of the adjacent aisles. (An alternative form of the portal method assumes that the wind load is distributed uniformly among the aisles, irrespective of their relative widths.)

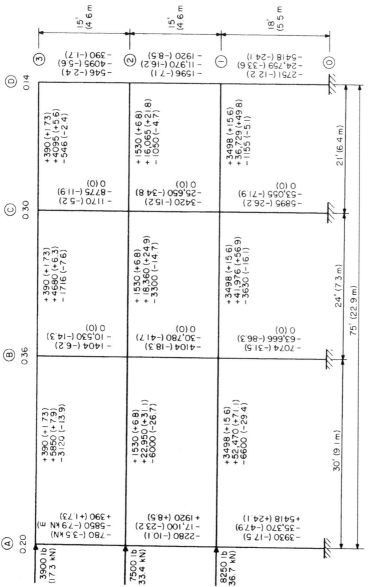

FIGURE 15. Wind-stress analysis by portal method.

In this analysis, we consider the *end moments* of a member, i.e., the moments exerted at the ends of the member by the joints. The sign conventions used are as follows. An end moment is positive if it is clockwise. The shear is positive if the lateral forces exerted on the member by the joints constitute a couple having a counterclockwise moment. An axial force is positive if it is tensile.

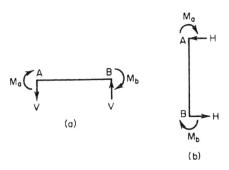

FIGURE 16

Figure 16a and b represents a beam and column, respectively, having positive end moments and positive shear. By applying the second assumption, $M_a = M_b = M$, Eq. a; $V = 2M/L$, or $M = VL/2$, Eq. b; $H = 2M/L$, or $M = HL/2$, Eq. c. In Fig. 15, the calculated data for each member are recorded in the order indicated.

The shear factor equals the ratio of the average width of the adjacent aisles to the total width. Or, line A, $15/75 = 0.20$; line B, $(15 + 12)/75 = 0.36$; line C, $(12 + 10.5)/75 = 0.30$; line D, $10.5/75 = 0.14$. For convenience, record these values in Fig. 15.

2. Compute the shear in each column
For instance, column A-2-3, $H = -3900(0.20) = -780$ lb (-3.5 kN); column C-1-2, $H = -(3900 + 7500)0.30 = -3420$ lb (-15.2 kN).

3. Compute the end moments of each column
Apply Eq. c. For instance, column A-2-3, $M = \frac{1}{2}(-780)15 = -5850$ ft·lb (-7932.6 N·m); column D-0-1, $M = \frac{1}{2}(-2751)18 = -24,759$ ft·lb ($-33,573.2$ N·m).

4. Compute the end moments of each beam
Do this by equating the algebraic sum of end moments at each joint to zero. For instance, at line 3: $M_{AB} = 5850$ ft·lb (7932.6 N·m); $M_{BC} = -5850 + 10,530 = 4680$ ft·lb (6346.1 N·m); $M_{CD} = -4680 + 8775 = 4095$ ft·lb (5552.8 N·m). At line 2: $M_{AB} = 5850 + 17,100 = 22,950$ ft·lb (31,120.2 N·m); $M_{BC} = -22,950 + 30,780 + 10,530 = 18,360$ ft·lb (24,896.0 N·m).

5. Compute the shear in each beam
Do this by applying Eq. b. For instance, beam B-2-C, $V = 2(18,360)724 = 1530$ lb (6.8 kN).

6. Compute the axial force in each member
Do this by drawing free-body diagrams of the joints and applying the equations of equilibrium. It is found that the axial forces in the interior columns are zero. This condition stems from the first assumption underlying the portal method and the fact that each interior column functions as both the leeward column of one portal and the windward column of the adjacent portal.

The absence of axial forces in the interior columns in turn results in the equality of the shear in the beams at each tier. Thus, the calculations associated with the portal method of analysis are completely self-checking.

WIND-STRESS ANALYSIS BY CANTILEVER METHOD

For the bent in Fig. 17, calculate all shears, end moments, and axial forces induced by the wind loads by applying the cantilever method of wind-stress analysis. For this purpose, assume that the columns have equal cross-sectional areas.

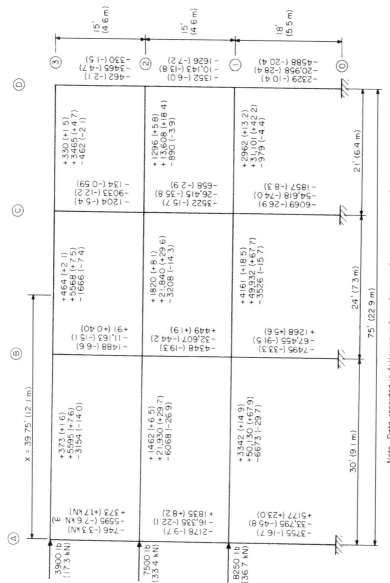

FIGURE 17. Wind-stress analysis by cantilever method.

Calculation Procedure:

1. Compute the shear and moment on the bent at midheight of each horizontal row of columns

The cantilever method, which is somewhat more rational than the portal method, considers that the bent behaves as a vertical cantilever. Consequently, the direct stress in a column is directly proportional to the distance from the column to the centroid of the combined column area. As in the portal method, the assumption is made that the point of contraflexure in each member lies at its center. Refer to the previous calculation procedure for the sign convention.

Computing the shear and moment on the bent at midheight, we have the following. Upper row: $H = 3900$ lb (17.3 kN); $M = 3900(7.5) = 29{,}250$ ft·lb (39,663.0 N·m). Center row: $H = 3900 + 7500 = 11{,}400$ lb (50.7 kN); $M = 3900(22.5) + 7500(7.5) = 144{,}000$ ft·lb (195.3 kN·m). Lower row: $H = 11{,}400 + 8250 = 19{,}650$ lb (87.5 kN); $M = 3900(39) + 7500(24) + 8250(9) = 406{,}400$ ft·lb (551.1 kN·m), or $M = 144{,}000 + 11{,}400(16.5) + 8250(9) = 406{,}400$ ft·lb (551.1 kN·m), as before.

2. Locate the centroidal axis of the combined column area, and compute the moment of inertia of the area with respect to this axis

Take the area of one column as a unit. Then $x = (30 + 54 + 75)/4 = 39.75$ ft (12.12 m); $I = 39.75^2 + 9.75^2 + 14.25^2 + 35.25^2 = 3121$ sq.ft. (289.95 m²).

3. Compute the axial force in each column

Use the equation $f = My/I$. The y/I values are

	A	B	C	D
y	39.75	9.75	−14.25	−35.25
y/I	0.01274	0.00312	−0.00457	−0.01129

Then column A-2-3, $P = 29{,}250(0.01274) = 373$ kips (1659 kN); column B-0-1, $P = 406{,}400(0.00312) = 1268$ kips (5640 kN).

4. Compute the shear in each beam by analyzing each joint as a free body

Thus, beam A-3-B, $V = 373$ lb (1659 N); beam B-3-C, $V = 373 + 91 = 464$ lb (2.1 kN); beam C-3-D, $V = 464 − 134 = 330$ lb (1468 N); beam A-2-B, $V = 1835 − 373 = 1462$ lb (6.5 kN); beam B-2-C, $V = 1462 + 449 − 91 = 1820$ lb (8.1 kN).

5. Compute the end moments of each beam

Apply Eq. b of the previous calculation procedure. Or for beam A-3-B, $M = \frac{1}{2}(373)(30) = 5595$ ft·lb (7586.8 N·m).

6. Compute the end moments of each column

Do this by equating the algebraic sum of the end moments at each joint to zero.

7. Compute the shear in each column

Apply Eq. c of the previous calculation procedure. The sum of the shears in each horizontal row of columns should equal the wind load above that plane. For instance, for the center row, $\Sigma H = −(2178 + 4348 + 3522 + 1352) = −11{,}400$ lb (−50.7 kN), which is correct.

HANGERS, CONNECTORS, AND WIND-STRESS ANALYSIS

8. Compute the axial force in each beam by analyzing each joint as a free body

Thus, beam A-3-B, $P = -3900 + 746 = -3154$ lb (-14.0 kN); beam B-3-C, $P = -3154 + 1488 = -1666$ lb (-7.4 kN).

WIND-STRESS ANALYSIS BY SLOPE-DEFLECTION METHOD

Analyze the bent in Fig. 18*a* by the slope-deflection method. The moment of inertia of each member is shown in the drawing.

Calculation Procedure:

1. Compute the end rotations caused by the applied moments and forces; superpose the rotation caused by the transverse displacement

This method of analysis has not been applied extensively in the past because the arithmetic calculations involved become voluminous where the bent contains many joints. However, the increasing use of computers in structural design is overcoming this obstacle and stimulating a renewed interest in the method.

Figure 19 is the elastic curve of a member subjected to moments and transverse forces applied solely at its ends. The sign convention is as follows: an end moment is positive if it is clockwise; an angular displacement is positive if the rotation is clockwise; the transverse displacement Δ is positive if it rotates the member in a clockwise direction.

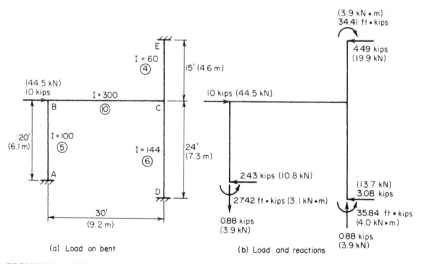

FIGURE 18. (*a*) Load on bent; (*b*) load and reactions.

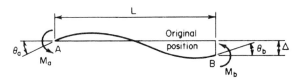

FIGURE 19. Elastic curve of beam.

Computing the end rotations gives $\theta_a = (L/6EI)(2M_a - M_b) + \Delta/L$; $\theta_b = (L/6EI)(-M_a + 2M_b) + \Delta/L$. These results may be obtained by applying the moment-area method or unit-load method given in Sec. 1.

2. Solve the foregoing equations for the end moments
Thus,

$$M_a = \left(\frac{2EI}{L}\right)\left(2\theta_a + \theta_b - \frac{3\Delta}{L}\right) \quad M_b = \left(\frac{2EI}{L}\right)\left(\theta_a + 2\theta_b - \frac{3\Delta}{L}\right) \quad (4)$$

These are the basic slope-deflection equations.

3. Compute the value of I/L for each member of the bent
Let K denote this value, which represents the relative stiffness of the member. Thus $K_{ab} = 100/20 = 5$; $K_{cd} = 144/24 = 6$; $K_{be} = 300/30 = 10$; $K_{ce} = 60/15 = 4$. These values are recorded in circles in Fig. 18.

4. Apply Eq. 4 to each joint in turn
When the wind load is applied, the bent will deform until the horizontal reactions at the supports total 10 kips (44.5 kN). It is evident, therefore, that the end moments of a member are functions of the *relative* rather than the absolute stiffness of that member. Therefore, in writing the moment equations, the coefficient $2EI/L$ may be replaced with I/L; to view this in another manner, $E = \frac{1}{2}$.

Disregard the deformation associated with axial forces in the members, and assume that joints B and C remain in a horizontal line. The symbol M_{ab} denotes the moment exerted on member AB at joint A. Thus $M_{ab} = 5(\theta_b - 3\Delta/20) = 5\theta_b - 0.75\Delta$; $M_{dc} = 6(\theta_c - 3\Delta/24) = 6\theta_c - 0.75\Delta$; $M_{ec} = 4(\theta_c + 3\Delta/15) = 4\theta_c + 0.80\Delta$; $M_{ba} = 5(2\theta_b - 3\Delta/20) = 10\theta_b - 0.75\Delta$; $M_{cd} = 6(2\theta_c - 3\Delta/24) = 12\theta_c - 0.75\Delta$; $M_{ce} = 4(2\theta_c + 3\Delta/15) = 8\theta_c + 0.80\Delta$; $M_{cd} = 10(2\theta_b + \theta_c) = 20\theta_b + 10\theta_c$; $M_{cb} = 10(\theta_b + 2\theta_c) = 10\theta_b + 20\theta_c$.

5. Write the equations of equilibrium for the joints and for the bent
Thus, joint B, $M_{ba} + M_{bc} = 0$, Eq. a; joint C, $M_{cb} + M_{cd} + M_{ce} = 0$, Eq. b. Let H denote the horizontal reaction at a given support. Consider a horizontal force positive if directed toward the right. Then $H_a + H_d + H_e + 10 = 0$, Eq. c.

6. Express the horizontal reactions in terms of the end moments
Rewrite Eq. c. Or, $(M_{ab} + M_{ba})/20 + (M_{dc} + M_{cd})/24 - (M_{ec} + M_{ce})/15 + 10 = 0$, or $6M_{ab} + 6M_{ba} + 5M_{dc} + 5M_{cd} - 8M_{ec} - 8M_{ce} = -1200$, Eq. c'.

7. Rewrite Eqs. a, b, and c' by replacing the end moments with the expressions obtained in step 4
Thus, $30\theta_b + 10\theta_c - 0.75\Delta = 0$, Eq. A; $10\theta_b + 40\theta_c + 0.05\Delta = 0$, Eq. B; $90\theta_b - 6\theta_c - 29.30\Delta = -1200$, Eq. C.

8. Solve the simultaneous equations in step 7 to obtain the relative values of θ_b, θ_c and Δ

Thus $\theta_b = 1.244$; $\theta_c = -0.367$; $\Delta = 44.85$.

9. Apply the results in step 8 to evaluate the end moments

The values, in foot-kips, are: $M_{ab} = -27.42$ (-37.18 kN·m); $M_{dc} = -35.84$ (-48.6 kN·m); $M_{ec} = 34.41$ (46.66 kN·m); $M_{ba} = -21.20$ (-28.75 kN·m); $M_{cd} = -38.04$ (-51.58 kN·m); $M_{ce} = 32.94$ (44.67 kN·m); $M_{bc} = 21.21$ (28.76 kN·m); $M_{cb} = 5.10$ (6.92 kN·m).

10. Compute the shear in each member by analyzing the member as a free body

The shear is positive if the transverse forces exert a counterclockwise moment. Thus $H_{ab} = (M_{ab} + M_{ba})/20 = -2.43$ kips (-10.8 kN); $H_{cd} = -3.08$ kips (-13.7 kN); $H_{ce} = 4.49$ kips (19.9 kN); $V_{bc} = 0.88$ kip (3.9 kN).

11. Compute the axial force in AB and BC

Thus $P_{ab} = 0.88$ kip (3.91 kN); $P_{bc} = -7.57$ kips (-33.7 kN). The axial forces in *EC* and *CD* are found by equating the elongation of one to the contraction of the other.

12. Check the bent for equilibrium

The forces and moments acting on the structure are shown in Fig. 18*b*. The three equations of equilibrium are satisfied.

WIND DRIFT OF A BUILDING

Figure 20*a* is the partial elevation of the steel framing of a skyscraper. The wind shear directly above line 11 is 40 kips (177.9 kN), and the wind force applied at lines 11 and 12 is 4 kips (17.8 kN) each. The members represented by solid lines have the moments of inertia shown in Table 1, and the structure is to be analyzed for wind stress by the portal method. Compute the wind drift for the bent bounded by lines 11 and 12; that is, find the horizontal displacement of the joints on line 11 relative to those on line 12 as a result of wind.

Calculation Procedure:

1. Using the portal method of wind-stress analysis, compute the shear in each column caused by the unit loads

Apply the unit-load method presented in Sec. 1. For this purpose, consider that unit horizontal loads are applied to the structure in the manner shown in Fig. 20*b*.

The results obtained in steps 1, 2, and 3 below are recorded in Fig. 20*b*. To apply the portal method of wind-stress analysis, see the fourteenth calculation procedure in this section.

2. Compute the end moments of each column caused by the unit loads

3. Equate the algebraic sum of end moments at each joint to zero; from this find the end moments of the beams caused by the unit loads

4. Find the end moments of each column

Multiply the results obtained in step 2 by the wind shear in each panel to find the end moments of each column in Fig. 20*a*. For instance, the end moments of column C-11-12 are $-1.95(44) = -85.8$ ft·kips (-116.3 kN·m). Record the result in Fig. 20*a*.

1.170 STRUCTURAL ENGINEERING

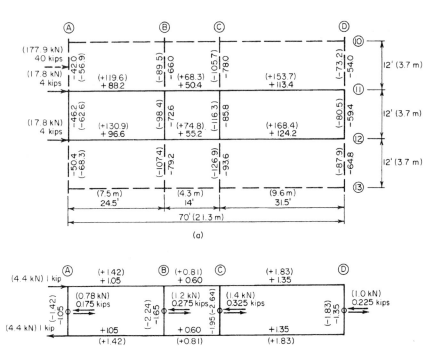

FIGURE 20

TABLE 1. Calculation of Wind Drift

Member	I, in⁴ (cm⁴)	L, ft (in.)	M_e, ft-kips (kN·m)	m_e, ft-kips (kN·m)	$M_e m_e L/I$
A-11-12	1,500 (62,430)	12 (3.66)	46.2 (62.6)	1.05 (1.42)	0.39
B-11-12	1,460 (60,765)	12 (3.66)	72.6 (98.5)	1.65 (2.24)	0.98
C-11-12	1,800 (74,916)	12 (3.66)	85.8 (116.3)	1.95 (2.64)	1.12
D-11-12	2,000 (83,240)	12 (3.66)	59.4 (80.6)	1.35 (1.83)	0.48
A-12-B	660 (27,469)	24.5 (7.47)	88.2 (119.6)	1.05 (1.42)	3.44
B-12-C	300 (12,486)	14 (4.27)	50.4 (68.3)	0.60 (0.81)	1.41
C-12-D	1,400 (58,268)	31.5 (9.60)	113.4 (153.8)	1.35 (1.83)	3.44
A-12-B	750 (31,213)	24.5 (7.47)	96.6 (130.9)	1.05 (1.42)	3.31
B-12-C	400 (16,648)	14 (4.27)	55.2 (74.9)	0.60 (0.81)	1.16
C-12-D	1,500 (62,430)	31.5 (9.60)	124.2 (168.4)	1.35 (1.83)	3.52
Total					19.25

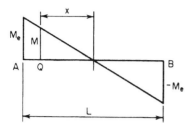
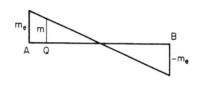

FIGURE 21. Bending-moment diagrams.

5. Find the end moments of the beams caused by the true loads
Equate the algebraic sum of end moments at each joint to zero to find the end moments of the beams caused by the true loads.

6. State the equation for wind drift
In Fig. 21, M_e and m_e denote the end moments caused by the true load and unit load, respectively. Then the

$$\text{Wind drift } \Delta = \frac{\sum M_e m_e L}{3EI} \qquad (5)$$

7. Compute the wind drift by completing Table 1
In recording end moments, algebraic signs may be disregarded because the product $M_e m_e$ is always positive. Taking the total of the last column in Table 1, we find $\Delta = 19.25(12)^2/[3(29)(10)^3] = 0.382$ in. (9.7 mm). For dimensional homogeneity, the left side of Eq. 5 must be multiplied by 1 kip (4.45 kN). The product represents the external work performed by the unit loads.

REDUCTION IN WIND DRIFT BY USING DIAGONAL BRACING

With reference to the previous calculation procedure, assume that the wind drift of the bent is to be restricted to 0.20 in. (5.1 mm) by introducing diagonal bracing between lines B and C. Design the bracing, using the gross area of the member.

Calculation Procedure:

1. State the change in length of the brace
The bent will be reinforced against lateral deflection by a pair of diagonal cross braces, each brace being assumed to act solely as a tension member. Select the lightest single-angle member that will satisfy the stiffness requirements; then compute the wind drift of the reinforced bent.

Assume that the bent in Fig. 22 is deformed in such a manner that B is displaced a horizontal

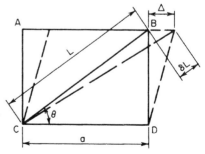

FIGURE 22

distance A relative to D. Let A = cross-sectional area of member CB; P = axial force in CB; P_h = horizontal component of P; δL = change in length of CB. From the geometry of Fig. 22, $\delta L = \Delta \cos \theta = a\Delta/L$ approximately.

2. Express P_h in terms of Δ
Thus, $P = aAE\Delta/L^2$; $P_h = P \cos \theta = Pa/L$; then

$$P_h = \frac{a^2 AE\Delta}{L^3} \tag{6}$$

3. Select a trial size for the diagonal bracing; compute the tensile capacity
A section of the AISC *Specification* limits the slenderness ratio for bracing members in tension to 300, and another section provides an allowable stress of 22 kips/sq.in. (151.7 MPa). Thus, $L^2 = 14^2 + 12^2 = 340$ sq.ft. (31.6 sq.in.); $L = 18.4$ ft (5.61 m); $r_{min} = (18.4 \times 12)/300 = 0.74$ in. (18.8 mm).

Try a $4 \times 4 \times \frac{1}{4}$ in. (101.6 × 101.6 × 6.35 mm) angle; $r = 0.79$ in. (20.1 mm); $A = 1.94$ sq.in. (12.52 cm^2); $P_{max} = 1.94(22) = 42.7$ kips (189.9 kN).

4. Compute the wind drift if the assumed size of bracing is used
By Eq. 6, $P_h = \{196/[(340)(18.4)(12)]\}\ 1.94(29)(10)^3\Delta = 147\Delta$ kips (653.9Δ N). The wind shear resisted by the columns of the bent is reduced by P_h, and the wind drift is reduced proportionately.

From the previous calculation procedure, the following values are obtained: without diagonal bracing, $\Delta = 0.382$ in. (9.7 mm); with diagonal bracing, $\Delta = 0.382/(44 - P_h)/44 = 0.382 - 1.28\Delta$. Solving gives $\Delta = 0.168 < 0.20$ in. (5.1 mm), which is acceptable.

5. Check the axial force in the brace
Thus, $P_h = 147(0.168) = 24.7$ kips (109.9 kN); $P = P_h L/a = 24.7(18.4)/14 = 32.5 < 42.7$ kips (189.9 kN), which is satisfactory. Therefore, the assumed size of the member is satisfactory.

LIGHT-GAGE STEEL BEAM WITH UNSTIFFENED FLANGE

A beam of light-gage cold-formed steel consists of two $7 \times 1\frac{1}{2}$ in. (177.8 × 38.1 mm) by no. 12 gage channels connected back to back to form an I section. The beam is simply supported on a 16-ft (4.88-m) span, has continuous lateral support, and carries a total dead load of 50 lb/lin ft (730 N/m). The live-load deflection is restricted to 1/360 of the span. If the yield-point stress f_y is 33,000 lb/sq.in. (227.5 MPa), compute the allowable unit live load for this member.

Calculation Procedure:

1. Record the relevant properties of the section
Apply the AISC *Specification for the Design of Light Gage Cold-Formed Steel Structural Members*. This is given in the AISC publication *Light Gage Cold-Formed Steel Design Manual*. Use the same notational system, except denote the flat width of an element by g rather than w.

The publication mentioned above provides a basic design stress of 20,000 lb/sq.in. (137.9 MPa) for this grade of steel. However, since the compression flange of the given member is unstiffened in accordance with the definition in one section of the publication, it may be necessary to reduce the allowable compressive stress. A table in the *Manual* gives the dimensions, design properties, and allowable stress of each section, but the allowable stress will be computed independently in this calculation procedure.

Let V = maximum vertical shear; M = maximum bending moment; w = unit load; f_b = basic design stress; f_c = allowable bending stress in compression; v = shearing stress; Δ = maximum deflection.

Record the relevant properties of the section as shown in Fig. 23: I_x = 12.4 in^4 (516.1 cm^4); S_x = 3.54 in^3 (58.0 cm^3); R = 3/16 in. (4.8 mm).

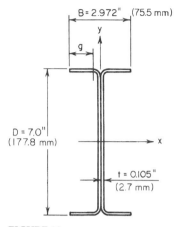

FIGURE 23

2. Compute f_c

Thus, $g = B/2 - t - R = 1.1935$ in. (30.3 mm); $g/t = 1.1935/0.105 = 11.4$. From the *Manual*, the allowable stress corresponding to this ratio is $f_c = 1.667 f_b - 8640 - 1(f_b - 12{,}950)g/t]/15 = 1.667(20{,}000) - 8640 - (20{,}000 - 12{,}950)11.4/15 = 19{,}340$ lb/sq.in. (133.3 MPa).

3. Compute the allowable unit live load if flexure is the sole criterion

Thus $M = f_c S_x = 19{,}340(3.54)/12 = 5700$ ft·lb (7729.2 N·m); $w = 8M/L^2 = 8(5700)/16^2 = 178$ lb/lin ft (2.6 kN/m); $w_{LL} = 178 - 50 = 128$ lb/lin ft (1.87 kN/m).

4. Investigate the deflection under the computed live load

Using $E = 29{,}500{,}000$ lb/sq.in. (203,373 MPa) as given in the AISC *Manual*, we have $\Delta_{LL} = 5 w_{LL} L^4/(384 E I_x) = 5(128)(16)^4 (12)^3/[384(29.5)(10)^6 12.4] = 0.516$ in. (13.1 mm); $\Delta_{LL,allow} = 16(12)/360 = 0.533$ in. (13.5 mm), which is satisfactory.

5. Investigate the shearing stress under the computed total load

Refer to the AISC *Specification*. For the individual channel, $h = D - 2t = 6.79$ in. (172.5 mm); $h/t = 64.7$; $64{,}000{,}000/64.7^2 > \tfrac{2}{3} f_b$; therefore, $v_{allow} = 13{,}330$ lb/sq.in. (91.9 MPa); the web area = $0.105(6.79) = 0.713$ sq.in. (4.6 cm^2); $V = \tfrac{1}{4}(178)16 = 712$ lb (3.2 kN); $v = 712/0.713 < v_{allow}$, which is satisfactory. The allowable unit live load is therefore 128 lb/lin ft (1.87 kN/m).

LIGHT-GAGE STEEL BEAM WITH STIFFENED COMPRESSION FLANGE

A beam of light-gage cold-formed steel has a hat cross section 8 × 12 in. (203.2 × 304.8 mm) of no. 12 gage, as shown in Fig. 24. The beam is simply supported on a span of 13 ft (3.96 m). If the yield-point stress is 33,000 lb/sq.in. (227.5 MPa), compute the allowable unit load for this member and the corresponding deflection.

Calculation Procedure:

1. Record the relevant properties of the entire cross-sectional area

Refer to the AISC *Specification* and *Manual*. The allowable load is considered to be the ultimate load that the member will carry divided by a load factor of 1.65. At ultimate load, the bending stress varies considerably across the compression flange. To surmount the

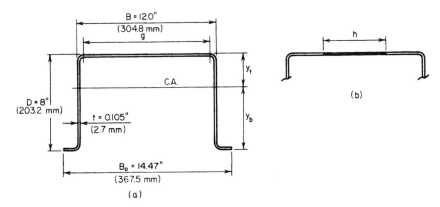

FIGURE 24

difficulty that this condition introduces, the AISC *Specification* permits the designer to assume that the stress is uniform across an *effective flange width* to be established in the prescribed manner. The investigation is complicated by the fact that the effective flange width and the bending stress in compression are interdependent quantities, for the following reason. The effective width depends on the compressive stress; the compressive stress, which is less than the basic design stress, depends on the location of the neutral axis; the location of the neutral axis, in turn, depends on the effective width.

The beam deflection is also calculated by establishing an effective flange width. However, since the beam capacity is governed by stresses at the ultimate load and the beam deflection is governed by stresses at working load, the effective widths associated with these two quantities are unequal.

A table in the AISC *Manual* contains two design values that afford a direct solution to this problem. However, the values are computed independently here to demonstrate how they are obtained. The notational system presented in the previous calculation procedure is used, as well as A' = area of cross section exclusive of compression flange; H = static moment of cross-sectional area with respect to top of section; y_b and y_t = distance from centroidal axis of cross section to bottom and top of section, respectively.

We use the AISC *Manual* to determine the relevant properties of the entire cross-sectional area, as shown in Fig. 24: $A = 3.13$ sq.in. (20.2 cm^2); $y_b = 5.23$ in. (132.8 mm); $I_x = 26.8$ in^4 (1115.5 cm^4); $R = 3/16$ in. (4.8 mm).

2. Establish the value of f_c for load determination

Use the relation $(8040t^2/f_c^{0.5})\{1 - 2010/[(f_c^{0.5}g)/t]\} = (H/D)(f_c + f_b)/f_c - A'$. Substituting gives $g = B - 2(t + R) = 12.0 - 2(0.105 + 0.1875) = 11.415$ in. (289.9 mm); $g/t = 108.7$; $gt = 1.20$ sq.in. (7.74 cm^2); $A = 3.13 - 1.20 = 1.93$ sq.in. (12.45 cm^2); $y_t = 8.0 - 5.23 = 2.77$ in. (70.36 cm); $H = 3.13(2.77) = 8.670$ in^3 (142.1 cm^3). The foregoing equation then reduces to $(88.64/f_c^{0.5})(1 - 18.49/f_c^{0.5}) = 1.084(f_c + 20,000)/f_c - 1.93$. By successive approximations, $f_c = 14,800$ lb/sq.in. (102.0 MPa).

3. Compute the corresponding effective flange width for load determination in accordance with the AISC Manual

Thus, $b = (8040t/f_c^{0.5})1 - 2010/[(f_c^{0.5}g)/t] = (8040 \times 0.105/14,800^{0.5})[1 - 2010/(14,800^{0.5} \times 108.7)] = 5.885$ in. (149.5 mm).

4. Locate the centroidal axis of the cross section having this effective width; check the value of f_c

Refer to Fig. 24b. Thus $h = g - b = 11.415 - 5.885 = 5.530$ in. (140.5 mm); $ht = 0.581$ sq.in. (3.75 cm^2); $A = 3.13 - 0.581 = 2.549$ sq.in. (16.45 cm^2); $H = 8.670$ in^3 (142.1 cm^3); $y_t = 8.670/2.549 = 3.40$ in. (86.4 mm); $y_b = 4.60$ in. (116.8 mm); $f_c = y_t/y_b = 3.40(20,000)/4.60 = 14,800$ lb/sq.in. (102.0 MPa), which is satisfactory.

5. Compute the allowable load

The moment of inertia of the net section may be found by applying the value of the gross section and making the necessary corrections. Applying $S_x = I_x/y_b$, we get $I_x = 26.8 + 3.13(3.40 - 2.77)^2 - 0.581(3.40 - 0.053)^2 = 21.53$ in^4 (896.15 cm^4). Then $S_x = 21.53/4.60 = 4.68$ in^3 (76.69 cm^3). This value agrees with that recorded in the AISC *Manual*.

Then $M = f_b S_x = 20,000(4.68)/12 = 7800$ ft·lb (10,576 N·m); $w = 8M/L^2 = 8(7800)/13^2 = 369$ lb/lin ft (5.39 kN/m).

6. Establish the value of f_y for deflection determination

Apply $(10,320 t^2/f_c^{0.5})[1 - 2580/(f_c^{0.5} g/t)] = (H/D)(f_c + f_b)/f_c - A'$, or $(113.8/f_c^{0.5}) \times (1 - 23.74/f_c^{0.5}) = 1.084(f_c + 20,000)/f_c - 1.93$. By successive approximation, $f_c = 13,300$ lb/sq.in. (91.7 MPa).

7. Compute the corresponding effective flange width for deflection determination

Thus, $b = (10,320 t/f_c^{0.5})[1 - 2580/(f_c^{0.5} g/t)] = (10,320 \times 0.105/13,300^{0.5})[1 - 2580/(13,300^{0.5} \times 108.7)] = 7.462$ in. (189.5 mm).

8. Locate the centroidal axis of the cross section having this effective width; check the value of f_c

Thus $h = 11.415 - 7.462 = 3.953$ in. (100.4 mm); $ht = 0.415$ sq.in. (2.68 cm^2); $A = 313 - 0.415 = 2.715$ sq.in. (17.52 cm^2); $H = 8.670$ in^3 (142.1 cm^3); $y_t = 8.670/2.715 = 3.19$ in. (81.0 mm); $y_b = 4.81$ in. (122.2 mm); $f_c = (3.19/4.81)20,000 = 13,300$ lb/sq.in. (91.7 MPa), which is satisfactory.

9. Compute the deflection

For the net section, $I_x = 26.8 + 3.13(3.19 - 2.77)^2 - 0.415(3.19 - 0.053)^2 = 23.3$ in^4 (969.8 cm^4). This value agrees with that tabulated in the AISC *Manual*. The deflection is $\Delta = 5wL^4/(384 E I_x) = 5(369)(13)^4(12)^3/[384(29.5)(10)^6 23.3] = 0.345$ in. (8.8 mm).

Related Calculations. New stadiums for football and baseball teams feature unique civil engineering design approaches to steel beams, columns, and surface areas. Thus, the Arizona Cardinals' new football stadium in Glendale, Arizona, will have two Brunel trusses supporting the roof of the stadium.

The Brunel trusses also support two transparent retractable panels that permit open-air games in good weather. In inclement weather, the two panels can be moved together to enclose the roof of the stadium. When the roof panels are closed the entire stadium can be climate-controlled using the facilities' air-conditioning system.

To further simulate outdoor conditions for the playing field, the entire 100-yard-long (100 m) natural-grass field can be rolled outdoors. Then the natural grass can receive both sunlight and rain to help the grass grow in a normal way. The field weighs some 9500 tons. It is moved outdoors on 542 wheels and has built-in drainage and watering piping. Some 76 motor-driven steel wheels power the movement of the field into, and out of, the stadium.

When the playing field is moved out of the stadium, other uses that do not require grass turf can be made of the facility. Thus, concerts, circus performances, industry shows, conventions, and the like, can be held in the stadium while the turf field is outdoors.

This structure is one example of the advancing use of civil engineering to meet the requirements of today's growing population. A sports stadium is leading the way to innovative design concepts aimed at making people's lives safer and more enjoyable.

INDEX

Note: Page numbers followed by "ff." indicate that the discussion continues on following pages.

Activated sludge reactor, **8**.1 to **8**.8
   aeration tank balance, **8**.5, **8**.6
   biochemical oxygen demand (BOD), **8**.1 to **8**.3, **8**.44 ff.
   food to microorganism ratio, **8**.4
   hydraulic retention time, **8**.2
   oxygen requirements, **8**.3
   reactor volume, **8**.4
   return activate sludge, **8**.4
   sludge quantity wasted, **8**.2
   volume of suspended solids (VSS), **8**.2
   wasted-activated sludge, **8**.4 to **8**.7
Active earth pressure on retaining wall, **4**.36
Aerated grit chamber design, **8**.16 ff.
   air supply required, **8**.18
   grit chamber dimensions, **8**.18
   grit chamber volume, **8**.16
   quantity of grit expected, **8**.18
Aerobic digester design, **8**.12 to **8**.16
   daily volume of sludge, **8**.13
   digester volume, **8**.15
   oxygen and air requirements, **8**.14 to **8**.15
   required VSS reduction, **8**.14
   volume of digested sludge, **8**.14
Air-lift pumps, selection of, **7**.9 to **7**.11
   disadvantages of, **7**.11
Air testing of sewers, **7**.37
Allowable-stress design, **1**.38
Alternative proposals, cost comparisons of, **9**.12 to **9**.22
Altitude of star, **5**.11 to **5**.13
Anaerobic digester design, **8**.44
   BOD entering digester, **8**.44
   daily quantity of volatile solids, **8**.44

Anaerobic digester design (*Cont*):
   percent stabilization, **8**.44
   required volume and loading, **8**.45
   volume of methane produced, **8**.46
Analysis of business operations, **9**.34 to **9**.39
   project planning using CPM/PERT, **9**.34 ff.
Area of tract, **5**.4 to **5**.8
Astronomy, field, **5**.11 to **5**.13
Average-end-area method, **5**.10, **5**.11
Average-grade method, **5**.25 to **5**.28
Axial load:
   and bending, **1**.62
   deformation caused by, **1**.62
   notational system for, **1**.62
   in steel beam column, **1**.34
      in braced frame, **1**.49
   in steel hanger, **1**.62
   stress caused by impact load, **1**.66
   (see also Stress(es) and strain: axial)
Axial member, design load in, **1**.33
Axial shortening of loaded column, **1**.40
Azimuth of star, **5**.11 to **5**.13

Balanced design:
   of prestressed-concrete beam, **2**.51
   of reinforced-concrete beam, **2**.3, **2**.18 ff.
   of reinforced-concrete column, **2**.35
Basins, rapid-mix and flocculation, **8**.28
Beam(s):
   bearing plates for, **1**.59 to **1**.61
   bending stress in, **1**.95 ff.
      and axial force, **1**.13
      with intermittent lateral support, **1**.10
      jointly supporting a load, **1**.96

Beam(s), bending stress in (*Cont*):
    with reduced allowable stress, **1.**11
    in riveted plate girder, **1.**21
    in welded plate girder, **1.**19
    (see also Bending moment)
  biaxial bending, **1.**49 ff.
  column (see Beam column)
  compact and non-compact, **1.**39 ff.
  composite action of, **1.**57
  composite steel-and-concrete, **2.**86 ff.
    highway bridge, **5.**35 to **5.**38
  composite steel-and-timber, **1.**80 ff.
  compound, **1.**79
  concentrated load on, **1.**76
  concrete slab, **1.**57
  conjugate, **1.**91
  continuous, **1.**89 to **1.**93
    of prestressed concrete, **2.**73,
        5.32 to **5.**106
    of reinforced concrete, **2.**23 ff.
    of steel, **1.**17, **1.**129
  deflection of, **1.**89 to **1.**93
  design moment, **1.**44
  economic section of, **1.**10
  lateral torsional buckling, **1.**44
  lightest section to support load, **1.**47 ff.
  minor- and major-axis bending in, **1.**42
  on movable supports, **1.**78
  with moving loads, **1.**103 to **1.**112
  orientation of, **1.**47
  prestressed concrete (see Prestressed-
    concrete beam)
  reinforced concrete (see Reinforced-
    concrete beam)
  shape, properties of, **1.**9 ff.
  shear, in concrete:
    connectors for, **1.**57
    force, **1.**56
  shear, in steel:
    center of, **1.**83
    flow, **1.**82
    on yielding support, **1.**96
  shear and bending moment in, **1.**76 ff.
  shear strength, **1.**43 to **1.**47
    of beam web, **1.**82
  shearing stress in, **1.**82 ff.
  statically indeterminate, **1.**95 to **1.**98
  steel (see Steel beam)
  stiffener plates for, **1.**43

Beam(s) (*Cont*):
  strength, for minor- and major-axis
    bending, **1.**42
  theorem of three moments, **1.**97
  timber (see Timber beam)
  vertical shear in, **1.**78
  web, **1.**58
  web stiffeners for, **1.**43
  welded section, **1.**47
  wood (see Timber beam)
Beam column:
  axial load on, **1.**49
  in braced frame, **1.**49 ff.
  and compression member, **2.**31
  first- and second-order moments in,
    1.49
  flexure and compression combined, **1.**49
  pile group as, **1.**88
  soil prism as, **1.**87
  steel, **1.**127 ff.
Beam connection:
  eccentric load on, **1.**118 ff.
    on pile group, **1.**88
    on riveted connection, **1.**118
    on welded connection, **1.**121
  riveted moment, **1.**117
  semirigid, **1.**147
  of truss members, **1.**146
  welded flexible, **1.**151
  welded moment, **1.**154
  welded seated, **1.**153
Beam web, shear strength of, **1.**58
Bearing plate, for beam, **1.**59
Bending flat plate, **1.**85
Bending moment:
  in bridge truss, **1.**108
  in column footing, **2.**42 ff.
  in concrete girders in T-beam bridge,
    5.35
  diagram, **1.**78 ff.
    for beam column, **1.**34
    for combined footing, **2.**42 to **2.**45
    for steel beam, **1.**13 ff.
    for welded plate girder, **1.**22
  in prestressed-concrete beam, **2.**59
  in reinforced-concrete beam, **2.**4
    compression member, **2.**35
  in steel beam, **1.**75 ff.
    continuous, **1.**129

Bending moment, in steel beam (*Cont.*):
    equation for, **1.**97 ff.
    by moment distribution, **1.**76 ff.
    for most economic section, **1.**8
    and shear in, **1.**76 ff.
    on yielding support for, **1.**95
    and theorem of three moments, **1.**97 ff.
    in three-hinged arch, **1.**110
    in welded plate girder, **1.**19 ff.
Bending stress:
    and axial force, **1.**13
    and axial load, **1.**50 ff.
    in beam, **1.**75 ff.
    in curved member, **1.**87
    with intermittent lateral support, **1.**10
    jointly supporting a load, **1.**96
    in riveted plate girder, **1.**21
    in welded plate girder, **1.**22
    (see also Bending moment)
Benefit-cost analysis, **9.**33
Bernoulli's theorem, **6.**7 ff.
Biogas plants, **8.**51
Bolted splice design, **3.**10
Borda's formula for head loss, **6.**14
Boussinesq equation, **4.**6, **4.**7
Bracing, diagonal, **1.**171
Branching pipes, **6.**15 ff.
Bridge, design procedure for, **5.**40 ff.
Bridge(s), highway, **5.**32 to **5.**106
    composite steel-and-concrete beam, **2.**90 to **2.**94
    bridge, **5.**35
    concrete T-beam, **5.**32
    girder, design of, **5.**32
Bridge filter sizing, traveling grate, **8.**23 ff.
Bridge girder, prestressed concrete, design of, **5.**45 ff.
    truss, **1.**104, **5.**93 ff.
Bulkhead:
    anchored analysis of, **4.**16
    cantilever analysis of, **4.**17
    thrust on, **4.**16
Buoyancy, **6.**3 to **6.**5
Business operations, analysis of, **9.**34 to **9.**39
    project planning using CPM/PERT, **9.**34 ff.
Butt splice, **1.**15

Cantilever bulkhead analysis, **4.**17
Cantilever retaining wall, **2.**46 ff.
Cantilever wind-stress analysis, **1.**164
Capacity-reduction factor, **2.**4
Capital, recovery of, **9.**6
Cardboard, corrugated, in municipal waste, **4.**34
Cash flow calculations, **9.**9
Celestial sphere, **5.**12
Centrifugal pump(s), **6.**24 to **6.**50
    characteristic curves for series installation, **6.**24 ff.
    cost reduction for:
        energy consumption and loss, **6.**60 ff.
        maintenance for, **6.**36
    driver speed, **6.**30 ff.
    energy cost reduction, **6.**59
    as hydraulic turbines, **6.**52
        applications for, **6.**57
        cavitation potential in, **6.**52
        constant-speed curves, **6.**53
        converting turbine to pump conditions, **6.**52
        number of stages of, **6.**52
        performance and flow rate, **6.**53
        specific speed of, **6.**52
    maintenance costs, reducing, **6.**37
    materials and parts selection, **6.**36 ff.
    minimum safe flow, **6.**50
    optimal operating speed, **6.**32 ff.
    ratings table for, **6.**36
    reverse flow, prevention of, **6.**27
    for safety-service, **6.**57 ff.
    seals, mechanical, **6.**43
    selection of, **6.**36 ff.
        for any system, **6.**36 ff.
        best operating speed, **6.**32 ff.
        most common error in, **6.**57
        for reduced cost, **6.**60 ff.
        similarity or affinity laws and, **6.**30 ff.
    similarity or affinity laws, **6.**30 ff.
Centrifuge, solid-bowl, for dewatering, **8.**23 ff.
Channel:
    nonuniform flow in, **6.**17
    uniform flow in, **6.**16

I.4 INDEX

Chlorination for wastewater disinfection, **8.49**
  capacity requirements, **8.49**
  coliform reduction in effluent, **8.49**
Circular settling tank design, **8.8** to **8.10**
  number of tanks, **8.9**
  peak flow, **8.8**
  surface area, **8.8**
Column(s):
  base, **1.158**
    grillage type, **1.160** ff.
  reinforced-concrete (see Reinforced-concrete column)
  steel (see Steel column)
  timber, **3.6**, **3.7**
Combined bending and axial loading, **1.85**
Communicating vessels, discharge between, **6.20**
Complete bridge structure, design of, **5.32** ff.
Composite mechanisms, theorem of, **1.134**
Composite member, thermal effects in, **1.72** ff.
Composite steel-and-concrete beam, **2.86** ff.
  highway bridge, **5.54** to **5.58**
Composite steel-and-concrete column, **1.56**
Composite steel-and-timber beam, **1.80** ff.
Composition of soil, **4.2**
Compound shaft, analysis of, **1.74**
Compression index, **4.26**
Compression member design:
  by ultimate-strength method, **2.31** ff.
  by working-stress method, **2.35** ff.
Compression test:
  triaxial, **4.10**
  unconfined, **4.8** to **4.10**
Concrete, modulus of elasticity of, **1.55**
Concrete, prestressed (see Prestressed-concrete beam(s))
Concrete, reinforced (see Reinforced-concrete beam(s))
Concrete slab, composite action of, **1.57**
Conjugate-beam method, **1.91**

Connection:
  beam-to-column (see Beam connection)
  horizontal shear resisting, **1.18** ff.
  pipe joint, **1.116**
  riveted, **1.112** to **1.118**
  semirigid, **1.147**
  timber, **3.8** to **3.12**
  welded, **1.151** to **1.160**
Connectors, shear, **1.57** ff.
Contaminants, soil (see Soil: contaminated)
Continuity, equation of, **6.8**
Converse-Labarre equation, **4.29**
Copper wire, in municipal wastes, **4.34**, **4.35**
Cost, plant capital, estimation of, **9.66**
Cost vs. benefit analysis, **9.33**
Coulomb's theory, **4.13**
Cover plates:
  for highway girder, **5.32** ff.
  for plate girder, **1.19** ff.
  for rolled section, **1.19**
Critical depth of fluid flow, **6.17** ff.
Critical-path method (CPM) in project planning, **9.34** ff.
Culmination of star, **5.14**
Curve:
  circular, **5.14** to **5.18**
  compound, **5.18** to **5.20**
  double meridian, **5.4**, **5.5**
  horizontal, **5.18**
  sight, **5.31**
  transition, **5.20** to **5.25**
  vertical, **5.31**
Cylindrical shaft, torsion of, **1.74**

Darcy-Weisbach formula, **6.11**
Declination of star, **5.13**
Deflection:
  of beam, **1.89** to **1.93**
  of cantilever frame, **1.93** ff.
  conjugate-beam method, **1.91**
  double-integration method, **1.89**
  moment-area method, **1.90**
  under moving loads, **1.12**
  of prestressed-concrete beam, **2.59**

Deflection (*Cont.*):
 of reinforced-concrete beam, **2**.30 ff.
 unit-load method, **1**.92
 virtual, **1**.126
Deformation of built-up member, **1**.126, **1**.131
Degree of saturation, **4**.2
Departure of line, **5**.2
Depreciation and depletion, **9**.9 ff.
 declining-balance, **9**.12
 straight-line, **9**.10 ff.
 sum-of-the-digits, **9**.12
Depth factor, **3**.3
Design, of complete bridge structure, **5**.40 ff.
 double-T roof in prestressed concrete, **5**.93
 post-tensioned girder in prestressed concrete, **5**.40
Dewatering of sludge, **8**.19 to **8**.23
Digester, wastewater sludge:
 aerobic system design, **8**.12 to **8**.16
 anaerobic system design, **8**.44 ff.
Disinfection of wastewater, chlorinated, **8**.49
Displacement of truss joint, **1**.30
Distance:
 double meridian, **5**.4, **5**.5
 sight, **5**.31
Double-integration method, **1**.89
Double-T roof, design of, **5**.93 ff.
Drainage pump, flow through, **6**.9
Drawdown in gravity wells, **7**.1 to **7**.8
 recovery, **7**.6 to **7**.9
Dummy pile, **4**.33
Dupuit formula in gravity well analysis, **7**.2, **7**.4

Earth thrust:
 on bulkhead, **4**.16 to **4**.19
 on retaining wall, **4**.13
 on timbered trench, **4**.14 to **4**.16
Earthwork requirements, **5**.10
Economics, engineering (see Engineering economics)
Effluent (see Sanitary sewer system design)
Elastic design, **1**.122
Elasticity, modulus of, **1**.62 ff.

Electrodialysis, area and power requirements, **8**.55
Embankment, stability of, **4**.20 to **4**.24
Engineering economics, **9**.4 to **9**.67
 alternative proposals, **9**.12 to **9**.22
  annual cost, after-tax basis, **9**.19
  annual cost of asset, **9**.23
  annual-cost studies, **9**.15
  asset replacement, **9**.20 to **9**.22
  capitalized cost, **9**.8 to **9**.20
  cost and income, **9**.14
  minimum asset life, **9**.14
 analysis of business operations, **9**.39 to **9**.42
  project planning using CPM/PERT, **9**.34 ff.
 capital recovery, **9**.6
 depreciation and depletion, **9**.9 to **9**.12
  accelerated cost recovery, **9**.10
  declining-balance, **9**.12
  effects of inflation, **9**.24 to **9**.26
  sinking-fund, **9**.16 ff.
  straight-line, **9**.9 ff.
  sum-of-the-digits, **9**.12
 evaluation of investments, **9**.26 to **9**.33
  allocation of capital, **9**.27 to **9**.30
  benefit-cost analysis, **9**.33
  payback period, **9**.32
  premium worth method, **9**.26
 inflation, effects of, **9**.24 to **9**.26
  anticipated, **9**.25
  at constant rate, **9**.24
  on replacement cost, **9**.24
  at variable rate, **9**.24
 interest calculations, **9**.3 to **9**.9
  compound, **9**.4
  effective rate, **9**.6
  simple, **9**.4
 nonuniform series, **9**.7
 perpetuity determination, **9**.7
 present worth, **9**.5 ff.
  of continuous cash flow of uniform rate, **9**.8
  of single payment, **9**.5
  of uniform-gradient series, **9**.6
  of uniform series, **9**.6
 probability, **9**.42 ff.
 sinking fund, principal in, **9**.5
 sinking-fund deposit, **9**.5

Engineering economics (*Cont.*):
   statistics and probability, **9.42** ff.
      arithmetic mean and median,
         9.43 ff.
      decision making, **9.54**
      of failure, **9.57**
      forecasting with a Markov process,
         9.63
      Monte Carlo simulation of
         commercial activity, **9.59**
      normal distribution, **9.49**
      number of ways of assigning work,
         9.46
      Pascal distribution, **9.47**
      perpetuity determination, **9.7**
      Poisson distribution, **9.48**
      population mean, **9.53**
      standard deviation, **9.43** ff.
      standard deviation from regression
         line, **9.62**
      uniform series, **9.8** ff.
Enlargement of pipe, **6.13** ff.
Environmental pollution (see Pollution,
   environmental)
Equipotential line, **4.4**
Equivalent-beam method, **4.18**
Euler equation, for column strength,
   1.26
Evaluation of investments, **9.26** to **9.39**
Eyebar design, **1.144**

Fatigue loading, **1.38**
Field astronomy, **5.11** to **5.14**
Fixed-end moment, **1.97** ff.
Flexural analysis:
   allowable-stress design, **1.122** ff.
   load and resistance factor design, **1.8** to
      1.61
   ultimate-stress design, **2.4** ff.
   working-stress design, **2.18** to **2.30**
Flocculation and rapid-mix basin, **8.26**
Flow line of soil mass, **4.4**
Flow net, **4.4**
Flowing liquid, power of, **6.10**
Fluid flow, **6.5** to **6.22**
   (see also Pump(s) and piping systems)
Fluid mechanics, **6.2** to **6.29**
   Francis equation in fluid discharge,
      6.10 ff.

Fluid mechanics (*Cont.*):
   hydrostatics, **6.2** to **6.7**
   of incompressible fluids, **6.5** to **6.22**
   incompressible fluids, mechanics of, **6.5**
      to **6.22**
   in pipes, flow determination for,
      6.12
   power of flowing, **6.10**
   raindrop, velocity of, **6.21**
   specific energy of mass of,
      6.17 ff.
Footing:
   combined, **2.39** to **2.42**
   isolated square, **2.40**
   settlement of, **4.27**
   sizing by Housel's method, **4.28**
   stability of, **4.24**
Force, hydrostatic, **6.2** to **6.5**
Francis equation for fluid discharge,
   6.10 ff.
Freyssinet cables, **2.67**

General wedge theory, **4.14** to **4.16**
Girder(s):
   plate, **1.19** to **1.22**, **5.32** to **5.65**
   posttensioned, design of, **2.67**
   steel plate, **1.83** to **1.85**
   T-beam, **2.7** ff.
   wood-plywood, **3.4** to **3.6**
Graphical analysis:
   of pile group, **4.29**
Gravity wells, **7.1** to **7.9**
   applications for, **7.3**
   base pressure curve, **7.3**
   discharging type, **7.4** to **7.6**
"Green" products, in reducing pollution,
   4.36
Grillage, as column support, **1.160**
Grit chamber, aerated, **8.16** to
   8.18
   air supply required, **8.18**
   chamber dimensions, **8.18**
   chamber volume, **8.16**
   quantity of grit expected, **8.18**
Groundwater:
   and drawdown of gravity well, **7.2** to
      7.4, **7.17**
   and ground surface, **7.3**
Gusset plate analysis, **1.146**

Hanger, steel, **1.**145
Head (pumps and piping):
  capacity:
    rotary pump ranges, **6.**40
    variable-speed, **6.**39 ff.
  computation of, **6.**36
  curves:
    plotting for, **6.**44 ff.
    system head, **6.**24
    types of, **6.**27
  effect of change in, **6.**31
  head loss, **6.**13 ff., **7.**11 to **7.**16
    Borda's formula for, **6.**13
    from pipe enlargement, **6.**13 ff.
    table for, **6.**35
  lift vs. friction, **6.**44 ff.
  in parallel pumping, **6.**27
  pressure loss (see head loss, above)
  in pump selection, **6.**36
  for vapor free liquid, **6.**33 ff.
  weir, variation on, **6.**10 ff.
Highway(s):
  bridge, rain runoff, **5.**106
  bridge design, **5.**32 to **5.**106
  transition spiral, **5.**20 to **5.**24
  volume of earthwork, **5.**10
Hoop stress, **1.**70
Hot-liquid pumps, suction head in, **6.**61
Housel's method, **4.**28
Hydraulic gradient, **4.**3, **6.**16
  in quicksand conditions, **4.**3
Hydraulic jump, **6.**18 ff.
  and power loss, **6.**22
Hydraulic radius, **6.**13
Hydraulic similarity, **6.**22
Hydraulic turbines:
  centrifugal pumps as, **6.**52 ff.
    applications for, **6.**56
    cavitation in, **6.**55 ff.
    constant-speed curves, **6.**55
    converting turbine to pump conditions, **6.**53
    number of stages of, **6.**53
    performance and flow rate, **6.**54
    specific speed of, **6.**53
Hydro power, **6.**52 to **6.**68
  "clean" energy from, **6.**65 ff.
  DOE cost estimates for, **6.**63

Hydro power (*Cont.*):
  Francis turbine in, **6.**62 ff.
  generating capacity, **6.**62
  generator selection, **6.**55
  small-scale generating sites, **6.**62 to **6.**65
  tail-water level, **6.**63
  turbines:
    design of, **6.**53 ff.
    efficiency and load sharing, **6.**64
    Francis turbine, **6.**62
    performance, by type, **6.**65 ff.
    selection of, **6.**66
    tube and bulb type, **6.**64
Hydrocarbons, petroleum, cleanup of, **4.**42
Hydroelectric power (see Hydro power)
Hydropneumatic storage tank sizing, **6.**51
Hydrostatics, **6.**2 to **6.**7
  Archimedes principle, **6.**3
  buoyancy and flotation, **6.**2 to **6.**7
  hydrostatic force, **6.**2 to **6.**5
    on curved surface, **6.**5
    on plane surface, **6.**3, **6.**4
    pressure prism, **6.**4
Hydroturbines:
  designs for, **6.**53 ff.
  efficiency and load sharing, **6.**64
  Francis turbine, **6.**62
  performance characteristics, **6.**53
  in small generating sites, **6.**62 ff.
  tube and bulb type, **6.**64

Impact load, axial stress and, **1.**66
Incompressible fluids, mechanics of, **6.**5 to **6.**22
  Bernoulli's theorem, **6.**7 ff.
  equation of continuity, **6.**5
Inflation, effects of, **9.**24 ff.
  anticipated, **9.**25
  at constant rate, **9.**24
  on replacement cost, **9.**24
  at variable rate, **9.**24
Influence line:
  for bridge truss, **1.**103 to **1.**112
  for three-hinged arch, **1.**110
Interaction diagram, **2.**32
Investments, evaluation of, **9.**20 to **9.**39
Irrigation, solar-powered pumps in, **6.**69

Joist(s):
    prestressed-concrete, **2.89** ff.
    wood, bending stress in, **3.2**

Kern distances, **2.60**
Knee, **1.155** ff.
Krey *f*-circle method of analysis, **4.22** to **4.24**

Laminated wood beam, design of, **3.13**
Landfills, **4.26** ff.
    mining of, **4.26**
Lap joint, welded, **1.120**
Lap splice, **1.114**
Laplace equation, **4.4**, **4.5**
Lateral torsional buckling, **1.44**
Latitude of line, **5.2**
Leveling, differential, **5.8**
Light-gage steel beam:
    with stiffened flange, **1.173**
    with unstiffened steel flange, **1.172**
Linear transformation, principle of, **2.75** ff.
Liquid(s):
    fluid mechanics, **6.2** to **6.22**
    Francis equation in fluid discharge, **6.10**
    incompressible fluids, mechanics of, **6.5** to **6.22**
    in pipes, flow determination for, **6.12**
    power of flowing, **6.10**
    specific energy of mass of, **6.17** ff.
    viscosity, in pumps and piping systems, **6.6** ff.
Load and resistance factor design (LRFD), **1.8** to **1.61**
Looping pipes, discharge of, **6.14** ff.
LRFD (see Load and resistance factor design)

Magnel diagram, **2.60** ff.
Manning formula factor, **6.13** ff.
Markov process in sales forecasting, **9.63** ff.
Maximum available moment, in composite beam, **2.92**
Maxwell's theorem, **1.112**
Mechanism method of plastic design, **1.131**

Member(s), **1.75** to **1.88**
    axial, design load in, **1.33**
    composite, thermal effects in, **1.72** ff.
    compression, beam column and, **2.31**
    compression, bending moment for, **2.31**
    curved, bending stress in, **1.87**
    steel tension, **1.36**
    timber, **3.8**
    truss, **1.103** to **1.112**
    ultimate-strength design, **2.31** ff.
    working-stress design, **2.18** ff.
    (see also Beam(s))
Membrane, for electrodialysis, **8.55**
Meridian of observer, **5.12**, **5.14**
Methane in anaerobic digester, **8.46**
Method:
    allowable-stress design (ASD), **1.122** to **1.141**
    average-end-area, **5.10**
    average-grade, **5.25** to **5.28**
    cantilever, for wind-stress analysis, **1.164**
    conjugate-beam, **1.91**
    Hardy Cross network analysis, **7.15** ff.
    Housel's, **4.28**
    load and resistance factor design (LRFD), **1.8** to **1.61**
    prismoidal, **5.10**
    of slices, **4.20**
    slope-deflection, **1.89** to **1.93**
    Swedish, for slope stability analysis, **4.20** to **4.22**
    tangent-offset, **5.25** to **5.28**
    ultimate-strength design, **2.4** ff.
    working-stress design, **2.18** to **2.30**
Mining landfills, **4.26**
Modulus of elasticity:
    for composite steel-and-concrete column, **1.56**
    for concrete, **1.57**
    for steel, **1.62**
Modulus of rigidity, **1.74**
Mohr's circle of stress, **1.68**, **4.8** to **4.10**
Moisture content of soil, **4.2**
Moment:
    bending (see Bending moment)
    of inertia, **1.26** ff.
    on riveted connection, **1.21**
    second-order, **1.49**
Moment-area method, **1.90**

Moment distribution, **1**.99
Monod kinetics, **8**.2
Monte Carlo simulation, **9**.59
Moving-load system:
  on beam, **1**.103 ff.
  on bridge truss, **1**.108 to **1**.110
Municipal wastes, recycle profits in, **4**.34 to **4**.36
  benefits from, **4**.35
  landfill space and, **4**.35
  recyclable materials, **4**.34, **4**.35
    price increase of, **4**.34 to **4**.36
  waste collection programs, **4**.35

Nadir of observer, **5**.12
Neutral axis in composite beam, **1**.46
Neutral point, **1**.106
Newspapers, in municipal wastes, **4**.34
Nonuniform series, **9**.7

Oblique plane, stresses on, **1**.67
Orifice, flow through, **6**.8

Parabolic arc, **2**.71 ff.
  change in grade, **5**.29
  coordinates of, **2**.71
  location of station on, **5**.28
  plotting, **5**.25 to **5**.28
  of prestressed-concrete beam, **2**.72 ff.
  summit, **5**.29
Parallel pumping economics:
  characteristic curves for, **6**.27 ff.
  check-valve location, **6**.30
  number of pumps used, **6**.27 ff.
  operating point for, **6**.28
  potential energy savings from, **6**.27
  system-head curve, **6**.28
Pascal probability distribution, **9**.47
Passive earth pressure on retaining wall, **4**.36
Payback period of investments, **9**.32
Permeability of soil, **4**.4
PERT in project planning, **9**.34 ff.
Pile-driving formula, **4**.28
Pile group:
  as beam column, **1**.88
  capacity of friction piles, **4**.29
  under eccentric load, **1**.88
  load distribution in, **1**.88, **4**.30 to **4**.34

Pipe joint, **1**.116
Pipe(s) and piping, **6**.5 to **6**.22
  Bernoulli's theorem, application of, **6**.7
  Borda's formula for head loss in pipe, **6**.13
  clay, in sewer pipes, **7**.30
  drainage pump, **6**.9
  eductor, **7**.10
  enlargement, effect on head, **6**.9 ff.
  flow of water in pipes, **6**.12 ff.
  industrial pipeline diagram, **6**.37
  looping pipes, discharge of, **6**.14 ff.
  pipe fittings:
    resistance coefficients of, **6**.33
    resistance of, and valves, **6**.34
  pipe size, **6**.13 ff.
    evaluation of, in pump selection, **6**.36
    Manning formula for selection of, **6**.13
    notation for, **6**.5
    for water supply, **7**.12, **7**.17, **7**.25 to **7**.29
  sewer and storm-water, **7**.24 to **7**.36
  suction and discharge piping, **6**.13 ff.
  Venturi meter, flow through, **6**.8
  in water-supply systems, **7**.11 to **7**.21
  (see also Pump(s) and piping systems)
Plant, capital cost estimation, **9**.66
Plastic containers, in municipal waste, **4**.34
Plastic design of steel structures, **1**.122 to **1**.141
  beam column, **1**.34 ff.
  continuous beam, **1**.129
  definitions relating to, **1**.122
  mechanism method of, **1**.126
  rectangular frame, **1**.131 to **1**.134
  shape factors in, **1**.123
  static method of, **1**.124 to **1**.128
  (see also Structural steel engineering and design)
Plastic media trickling filter design, **8**.36 ff.
Plastic modulus, **1**.122
Plastic moment, **1**.122 ff.
Plastification, defined, **1**.122
Plate, bending of, **1**.84 ff.
Plate girder, **1**.13, **5**.32 ff.
Poisson probability distribution, **9**.48
Poisson's ratio, **1**.84 ff.

Pollution, environmental, **4.**34 ff.
  "green" products and, **4.**36
  hydro power, "clean" energy from, **6.**65 ff.
  landfill space, **4.**35
  recycle profit potential, **4.**34 to **4.**36
  waste, municipal, **4.**34 to **4.**36
  waste sites, contaminated, **4.**2
Polymer dilution/feed system sizing, **8.**29
  predictable properties of, **8.**36
  required rotational rate of distributor, **8.**35
  treatability constant, **8.**35
  typical dosing rates for trickling filters, **8.**41
Porosity of soil, **4.**2
Portal method of wind-stress analysis, **1.**162
Post-tensioned girder, design of, **5.**45 ff.
Power of flowing liquid, **6.**10
Present worth:
  of continuous cash flow of uniform rate, **9.**9
  of single payment, **9.**5
  of uniform series, **9.**6
Pressure, soil, **4.**6 to **4.**8
  under dam, **1.**87
Pressure center of hydrostatic force, **6.**3
Pressure prism of fluid, **6.**4
Pressure vessel:
  prestressed, **1.**70
  thick-walled, **1.**70
  thin-walled, **1.**69
Prestress-moment diagram, **2.**75 ff.
Prestressed-concrete beam(s), **2.**51 to **2.**92, **5.**40 ff.
  in balanced design, **2.**50
  bending moment in, **2.**51
  in bridges, **5.**32 ff.
  continuity moment in, **2.**73
  continuous, **2.**73
    with nonprestressed reinforcement, **2.**77 ff.
    reactions for, **2.**89
  deflection of, **2.**62
  girder, **5.**45 ff.
  guides in design of, **2.**59, **5.**32 ff.
  kern of, **2.**60
  linear transformation in, **2.**75 ff.

Prestressed-concrete beam(s) (*Cont.*):
  loads carried by, **2.**59
  Magnel diagram for, **2.**60
  notational system for, **2.**51
  posttensioned, design of, **2.**67 ff.
  posttensioning of, **2.**50
  prestress shear and moment, **2.**51
  pretensioned, design of, **2.**59
  pretensioning of, **2.**51, **2.**59
  radial forces in, **2.**72
  section moduli of, **2.**58
  shear in, **2.**50
  stress diagrams for, **2.**52 ff.
  tendons in, **2.**52, 2,55, **2.**57
  trajectory, force in, **2.**72
  web reinforcement of, **2.**63 ff.
Prestressed concrete bridge design, **5.**32 ff.
Prestressed-concrete joist, **2.**89
Prestressed-concrete stair slab, **2.**90
Principal axis, **1.**27
Principal plane, **1.**67 ff.
Principal stress, **1.**68
Prismoidal method for earthwork, **5.**10
Probability:
  of failure, **9.**57
  normal distribution, **9.**52 ff.
  Pascal distribution, **9.**47
  Poisson distribution, **9.**48
  standard deviation, **9.**43
  standard deviation from regression line, **9.**62
Product of inertia, **1.**27
Profit potential, from recycling municipal waste, **4.**34 to **4.**36
Project planning, **9.**34
Pump(s) and piping systems, **6.**24 to **6.**44
  affinity laws for, **6.**30 ff.
  analysis of characteristic curves, **6.**44 ff.
    different pipe sizes, **6.**44 ff.
    duplex pump capacities, **6.**43
    effect of wear, **6.**49
    significant friction loss and lift, **6.**44
  centrifugal (see Centrifugal pump(s))
  check-valve, to prevent reverse flow, **6.**30
  closed-cycle solar-powered system, **6.**69 ff.
  discharge flow rate, **6.**9

INDEX   I.11

Pump(s) and piping systems (*Cont.*):
  drainage pump, flow through, **6.9**
  duplex plunger type, **6.44** ff.
  exit loss, **6.13** ff.
  fittings, resistance coefficients of, **6.33**
  head, **6.44** ff.
    analysis of characteristic curves, **6.44** ff.
    Borda's formula for head loss, **6.14**
    curves, plotting for, **6.44** ff.
    Darcy-Weisbach formula for friction, **6.11**
    effect of change in, **6.31**
    head loss and pipe enlargement, **6.13** ff.
  hydraulic jump, **6.18** ff.
    power loss resulting from, **6.18** ff.
  as hydraulic turbines, **6.52** ff.
  hydropneumatic storage tank, sizing of, **6.51**
  Manning formula for pipe-size selection, **6.13**
  materials for pump parts, **6.36** ff.
  minimum safe flow, **6.50**
  mostly lift, little friction, **6.46**
  no lift, all friction head, **6.46**
  optimal operating speed, **6.32**
  parallel pumping economics, **6.27** ff.
    characteristic curves for, **6.27** ff.
    check-valve location, **6.30**
    number of pumps used, **6.29**
    operating point for, **6.29**
    potential energy savings from, **6.27**
    system-head curve for, **6.28**
  in parallel system, **6.27**
  pipe-size selection, **6.13**
  pump type, by specific speed, **6.32**
  selection for any system, **6.36** ff.
    capacity required, **6.36**
    characteristics of modern pumps, **6.40**
    characteristics with diameter varied, **6.43**
    class and type, **6.38**
    common error in, **6.57**
    composite rating chart, **6.41**
    liquid conditions, **6.37**
    rating table, **6.40**

Pump(s) and piping systems, selection for any system (*Cont.*):
    selection guide, **6.37** ff.
    suction and discharge piping arrangements, **6.8**, **6.33**
    total head, **6.33**
    variable-speed head capacity, **6.39**
  selection for reduced energy cost, **6.57**
    best efficiency point (BEP) for, **6.61**
    energy efficiency pump, **6.60**
    and series pump operation, **6.24**
    specific speed and, **6.60** ff.
    specifications for operation below BEP, **6.60**
  selection of materials and parts, **6.36**, **6.51**
  series installation analysis, **6.24** ff.
    characteristic curves for, **6.44**
    in reducing energy consumption, **6.43**
  seriesed curve, **6.24**
  similarity and affinity laws, **6.30** ff.
  for small hydro power installations, **6.62**
  solar-powered system, **6.69**
    applications of, **6.69**
    closed-cycle design, **6.69**
    gas-release rate in, **6.69**
    Rankine-cycle turbine in, **6.70**
    solar collectors used for, **6.69**
  specific speed, by pump type, **6.32**
  static suction lift, **6.34**
  suction and discharge piping arrangements, **6.33**
  system-head curve, **6.26** ff.
  table for head loss determination, **6.36**
  total head for vapor free liquid, **6.33**
  valves, resistance of, **6.32**
  variable-speed head capacity, **6.39**
  variation in, on weir, **6.10** ff.
  velocity, pressure, and potential, **6.10**

Quicksand conditions determination, **4.3**

Rain, runoff from bridge, **5.106**
Rainfall:
  imperviousness of various surfaces to, **7.25**
  raindrop, velocity of, **6.21**

Rainfall (*Cont.*):
   storm-water runoff, **7.24, 7.25, 7.28,
     7.33** to **7.36**
Rankine's theory, **4.11**
Rapid-mix and flocculation basin design,
   8.28
   volume and power requirements, **8.28**
     for flocculation, **8.29**
     for rapid-mix basin, **8.28**
Rebhann's theorem, **4.13**
Reciprocal deflections, theorem of, **1.112**
Recycle profit potentials, in municipal
   wastes, **4.34** to **4.36**
Recycling of municipal waste:
   benefits from, **4.35**
   landfill space and, **4.35**
   profit potential in, **4.34** to **4.36**
   types of material in:
     copper, **4.34, 4.35**
     corrugated cardboard, **4.34**
     newspapers, **4.34**
     plastics, **4.34**
     prices of, **4.34** to **4.36**
   waste collection programs, **4.35**
Redtenbacker's formula, **4.29**
Reinforced-concrete beam, **2.4** to **2.13**
   in balanced design, **2.6** ff.
   bond stress in, **2.13**
   with compression reinforcement,
     2.9 ff.
   continuous: deflection of, **2.30** ff.
     design of, **2.14** to **2.16**
   equations of, **2.4** ff.
   failure in, **2.4**
   minimum widths, **2.4**
   with one-way reinforcement, **2.14**
   of rectangular section, **2.7** ff.
   shearing stress in, **2.11** ff.
     alternative methods for computing,
       2.11
   of T section, **2.7** ff.
   transformed section of, **2.20** ff.
   with two-way reinforcement, **2.16** to
     2.18
   ultimate-strength design of, **2.6** ff.
   web reinforcement of, **2.11** to **2.13, 2.24**
     to **2.26**
   working-stress design of, **2.18** to
     2.32

Reinforced-concrete column,
   2.35 ff.
   in balanced design, **2.39**
   footing for, **2.40** ff.
   interaction diagram for, **2.32** to
     2.34
   ultimate-strength design of, **2.4** ff.
   working-stress design of, **2.18** ff.
Retaining wall:
   active and passive earth pressure on,
     4.36
   cantilever, design of, **2.46**
   earth thrust on, **4.11** to **4.16**
Reynolds number, **6.5** ff.
Right ascension of star, **5.12**
Rigidity, modulus of, **1.74**
Riveted connection(s), **1.112** to **1.118**
   capacity of rivet in, **1.113**
   eccentric load on, **1.118** ff.
   moment on, **1.113**
Roof, double-T, design of, **5.93** ff.
Rotary-lobe sludge pump sizing, **8.41**
   flow rate required, **8.41**
   head loss in piping system, **8.42**
     multiplication factor, **8.42**
   pump horsepower (kW) required,
     8.43
     installed horsepower (kW),
       8.44
   pump performance curve, **8.43**
   pump selection, **8.40**
Roughness coefficient, **6.13**
Route design, **5.1** to **5.39**

Sanitary sewer system design, **8.53** ff.
   design factors, **8.49**
   lateral sewer size, **8.50**
   Manning formula conveyance factor,
     8.47
   required size of main sewer, **8.48**
   sanitary sewage flow rate, **8.46**
   sewer size with infiltration, **8.48**
Series pumping, **6.24** ff.
   characteristic curves for, **6.44**
   in reducing energy consumption,
     6.22
   seriesed curve, **6.26** ff.
   system-head curve, **6.26**
Settling tank design, **8.8** to **8.10**

Sewage-treatment method selection,
  8.53 ff.
  biogas plants, **8.**51
  daily sewage flow rate, **8.**50
  industrial sewage equivalent, **8.**50
  typical efficiencies, **8.**51
  wet processes, **8.**53
Sewer, air testing of, **7.**37
Sewer systems (see Storm-water and sewer systems and Sanitary sewer system design)
Shape factor, **1.**18 ff.
Shear:
  in beam, **1.**75 ff.
  in bridge truss, **1.**106, **5.**50 to **5.**52
  in column footing, **2.**40 ff.
  in concrete slab, for composite action, **1.**157
  of prestressed concrete, **2.**53
  punching, **2.**41, **2.**46
  on riveted connection, **1.**118
  for welded connection, **1.**121 ff.
Shear center, **1.**83
Shear connectors, **1.**57
Shear diagram:
  for beam, **1.**75 ff.
  for combined footing, **2.**42
Shearing stress (see Stress(es) and strain: shearing stress)
Shrink-fit stress, **1.**73
Sight distance, **5.**31
Similarity, hydraulic, **6.**22
Sinking fund, **9.**5
Slenderness ratio, **1.**29
Slices, method of, **4.**20 to **4.**22
Slope, stability of:
  by f-circle method, **4.**22 to **4.**24
  method of slices, **4.**20 to **4.**22
Slope-deflection method of wind-stress analysis, **1.**167 ff.
Sludge, sanitary wastewater treatment of, **8.**1 to **8.**44
  activated sludge reactor design, **8.**1 to **8.**8
  aerated grit chamber design, **8.**16 to **8.**18
  aerobic digester design, **8.**12 to **8.**16
  anaerobic digester design, **8.**44 ff.
  rotary-lobe sludge pump sizing, **8.**41 ff.

Sludge, sanitary wastewater treatment of (*Cont.*):
  solid-bowl centrifuge for dewatering, **8.**23 ff.
  thickening of wasted-activated sludge, **8.**4 to **8.**7
Small hydropower sites, **6.**62 ff.
  "clean" energy from, **6.**65
  DOE operating-cost estimates, **6.**66
  efficiency falloff and load sharing, **6.**64
  Francis turbine in, **6.**62
  importance of tail-water level, **6.**63
  turbine design, **6.**52 ff.
  typical power-generating capacity, **6.**62
Soil:
  composition of, **4.**2
  compression index of, **4.**26
  consolidation of, **4.**25
  contaminated, **4.**2, **7.**33
    and water supply, **7.**23
  flow net in, **4.**4 to **4.**6
  moisture content of, **4.**2, **4.**3
  permeability of, **4.**4
  porosity of, **4.**2
  pressure (see Soil pressure)
  quicksand conditions, **4.**3
  shearing capacity of, **4.**8 to **4.**10
  specific weight of, **4.**3
  thrust on bulkhead, **4.**16
Soil mechanics, **4.**1 to **4.**37
  mining landfills, **4.**26
  municipal wastes, recycle profits in, **4.**34 to **4.**36
Soil pressure:
  caused by point load, **4.**6
  caused by rectangular loading, **4.**7
  under dam, **1.**87

Solar-powered pumps, **6.**69
  applications of, **6.**69
  closed-cycle design, **6.**69
  gas-release rate in, **6.**69
  Rankine-cycle turbine in, **6.**69
  refrigerant selection for, **6.**69
  solar collectors used for, **6.**69

Solid-bowl centrifuge for sludge
    dewatering, **8.**49 ff.
  capacity and number of centrifuges,
      8.19
  centrifugal force, **8.**22
  dewatered sludge cake discharge rate,
      8.21
  selecting number of centrifuges needed,
      8.19
  sludge feed rate required, **8.**20
  solids capture, **8.**15
Space frame, **1.**135
Spiral, transition, **5.**20 to **5.**25
Stability:
  of embankment, **4.**20 to **4.**24
  of footing, **4.**24
  of slope, **4.**20 to **4.**24
  of vessel, **6.**6
Stadia surveying, **5.**9
Stair slab, prestressed-concrete, **2.**90
Star, azimuth of, **5.**11 to **5.**13
  culmination of, **5.**14
Star strut, **1.**28
Statically indeterminate structures, **1.**95 to
    1.101
  beam(s), **1.**95 ff.
    bending moment of, **1.**96
    bending stress of beam, **1.**96
    theorem of three moments, **1.**97
    truss, analysis of, **1.**101
Statistics and probability, **9.**42 ff.
Steel beam(s), **1.**8 to **1.**28
  composite concrete and, **2.**84 ff.
  continuous, **1.**94 ff.
    elastic design of, **1.**122 ff.
    plastic design of, **1.**122
  with continuous lateral support, **1.**8
  cover-plated, **1.**22
  encased in concrete, **2.**84 ff.
  with intermittent lateral support, **1.**10
  light gage, **1.**172 ff.
  with reduced allowable stress, **1.**11 ff.
  shear in, **1.**76 ff.
  shearing stress in, **1.**76 ff.
  stiffener plates for, **1.**121
Steel column, **1.**27 to **1.**35
  axial shortening when loaded, **1.**33
  base for, **1.**158 ff.
  beam column, **1.**34 ff., **1.**49

Steel column (*Cont.*):
  built-up, **1.**27
  compressive strength, **1.**26
    of composite, **1.**55 ff.
    of welded section, **1.**27
  concrete-filled, **1.**55
  effective length of, **1.**29
  with end moments, **1.**34
  under fatigue loading, **1.**33
  with grillage support, **1.**160
  with intermediate loading, **1.**32
  lacing of, **1.**31
  with partial restraint, **1.**30
  of star-strut section, **1.**28
  with two effective lengths, **1.**29
  welded section, **1.**121
Steel hanger analysis, **1.**145
Steel structures (see Structural steel
    engineering and design)
Steel tension member, **1.**34
Stiffener plates, **1.**43, **1.**151 to **1.**154
Storage tank, hydropneumatic, **6.**51
Storm-water and sewer systems, **7.**24 to
    7.36
  runoff rate and rainfall intensity, **7.**24
    by area, **7.**25
    rational method, **7.**24
    by surface, **7.**24
    Talbot formulas for, **7.**24
    (see also Sanitary sewer system
      design)
  sewer pipes, **7.**25 to **7.**36
    bedding requirements of, **7.**29 to **7.**33
    capacities, **7.**34
    clay pipe strength, **7.**30
    earth load on, **7.**29 to **7.**33
    embedding method, selection of, **7.**31
    sanitary systems, **7.**28
    separate vs. combined design types,
      7.36
    sizing for flow rates, **7.**25 to **7.**29, **7.**33
    slope of, **7.**26, **7.**36
    typical plot plan, **7.**35
Stress(es) and strain, **1.**62 to **1.**74
  axial, **1.**62 to **1.**63, **1.**158
  bending, **1.**76 ff.
  bond, **2.**13 ff.
  in compound shaft, **1.**74
  in flexural members, **1.**75 ff.

Stress(es) and strain (*Cont.*):
  hoop, **1.**69
  moving loads, **1.**103 to **1.**112
  on oblique plane, **1.**67
  principal, **1.**68
  in rectangular beam, **2.**20
  shearing stress, **1.**62 ff.
    cylindrical, torsion of, **1.**74
    in homogeneous beam, **1.**80
    in prestressed cylinder, **1.**70
    in reinforced-concrete beam, **2.**9 ff.
    shrink-fit and radial pressure, **1.**73
    in steel beam, **1.**76 to **1.**83
    thermal, **1.**72 ff.
    in timber beam, **3.**2, **3.**3
Structural steel engineering and design, **1.**1 to **1.**175
  axial member, design load in, **1.**62
  beam connection:
    riveted moment, **1.**149 to **1.**151
    semirigid, **1.**151
    welded moment, **1.**154
    welded seated, **1.**153
  column base, for axial load, **1.**158
    for end moment, **1.**158
    grillage type, **1.**160
  composite steel-and-concrete beam, **1.**80, **2.**92
  bridge, **5.**35 ff.
  connection, beam-to-column:
    of truss members, **1.**146
  eccentric load, **1.**118
    on pile group. **1.**88
    on rectangular section, **1.**95
    on riveted connection, **1.**118
    on welded connection, **1.**121
  eyebar, **1.**144
  gusset plate, **1.** 146
  hanger, steel, **1.**145
  knee:
    curved, **1.**155 to **1.**157
    rectangular, **1.**155
  stair slab, **2.**90
  steel beam, **1.**172 to **1.**173
    encased in concrete, **2.**84 ff.
    light-gage, **1.**172 to **1.**173
  wind drift, **1.**169 to **1.**171
    reduction with diagonal bracing, **1.**171

Structural steel engineering and design (*Cont.*):
  wind-stress analysis, **1.**164 to **1.**167
    cantilever method, **1.**164
    portal method, **1.**162
    slope-deflection method, **1.**167
Surveying, **5.**1 to **5.**31
  field astronomy, **5.**11 to **5.**14
  land and highway, **5.**1 to **5.**31
  stadia, **5.**9
Swedish method for slope analysis, **4.**20 to **4.**22

T-beam reinforced-concrete, **1.**173, **2.**7 ff.
Tangent-offset method, **5.**28
Tangential deviation, **1.**90
Tank(s):
  aeration, in wastewater treatment, **8.**6
  circular, in wastewater treatment, **8.**8 to **8.**10
  hydropneumatic, **6.**57
Temperature reinforcement, **1.**173
Tension member, steel, **1.**26
Terzaghi general wedge theory, **4.**14
Terzaghi theory of consolidation, **4.**25
Thermal effects in structural members, **1.**71 ff.
Thermodynamics, first law of, **6.**35
Three moments, theorem of, **1.**97 ff.
Timber beam, **3.**1 to **3.**4
  bending stress in, **3.**2
  bolted splice, **3.**10
  composite steel and, **1.**45 to **1.**47
  depth factor of, **3.**3
  lateral load on nails in, **3.**9
  screw loads in, **3.**10
  shearing stress in, **3.**3
Timber column, **3.**6, **3.**7
Timber connection, **3.**11
Timber engineering, **3.**1 to **3.**13
Timber member under oblique force, **3.**7
Torsion of shaft, **1.**74
Tract:
  area of, **5.**5 to **5.**8
    irregular, **5.**7
    rectilinear, **5.**4
  partition of, **5.**5 to **5.**7

Trajectory:
   concordant, **2.**77
   linear transformation of, **2.**75 ff.
Transformed section, **1.**80
Transition spiral, **5.**20 to **5.**25
Traveling-grate bridge filter sizing, **8.**23 to **8.**25
Traverse, closed, **5.**2 to **5.**7
Trench, earth thrust on timbered, **4.**4 to **4.**16
Trickling filter design, **8.**29 to **8.**33
   BOD loading for first-stage filter, **8.**31
   filter efficiency, **8.**29
   plastic media type, **8.**33 to **8.**36
Truss:
   bridge, **1.**104, **5.**50
   influence line:
     for bending moment in, **1.**108
     with moving loads, **1.**106 ff.
     for shear in, **1.**104
     statically indeterminate, **1.**101
     by uniform loads, **1.**106
   joint, displacement of, **1.**65
Turbines:
   hydraulic:
     centrifugal pumps as, **6.**52 ff.
     converting turbine to pump conditions, **6.**52
   hydroturbines:
     designs for, **6.**52 to **6.**58
     efficiency and load sharing, **6.**52
     Francis turbine, **6.**62
     performance, by type, **6.**65
     in small-scale generating sites, **6.**62
     tube and bulb type, **6.**62 ff.
Turbulent flow, **6.**11 ff.
Two-way slab, **2.**16 ff.

Ultimate load, **1.**122
Ultimate-strength design:
   for compression members, **2.**31 ff.
   for flexural members, **2.**4 ff.
Uniform series, **9.**6 ff.
Unit-load method, **1.**92

Venturi meter, flow through, **6.**8 ff.
Vertical parabolic curve:
   containing given point, **5.**29

Vertical parabolic curve (*Cont.*):
   plotting of, **5.**25 to **5.**28
   sight distance on, **5.**31
Virtual displacements, theorem of, **1.**139
Visibility, on vertical curve, **5.**31
Void ratio of soil, **4.**26

Wall, retaining (see Retaining wall)
Waste:
   contaminated sites, **4.**34
   municipal, **4.**34
     incineration of, **4.**35
     landfill area required for, **4.**36
     rate of generation of, **4.**36
     recycle profit potential in, **4.**34 to **4.**36
Waste-activated sludge thickening, **8.**10 to **8.**12
   size of gravity belt thickener, **8.**11
   sludge and filtrate flow rates, **8.**11
   solids capture, **8.**12
Wastewater disinfection, chlorination system for, **8.**44
Wastewater treatment and control, **8.**1 to **8.**55
   (see also Sanitary sewer system design)
Water pollution:
   and hazardous wastes, **7.**23
   impurities in water-supply, **7.**21 to **7.**24
Water-supply systems, **7.**1 to **7.**21
   air-lift pump selection, **7.**9 to **7.**11
     compressor capacity, **7.**9
     submergence, effect of, **7.**10
   choice of pipe for, **7.**11
   demand curve for typical week, **7.**4
   fire safety requirements, **7.**12 ff.
   flow rate and pressure loss, **7.**12 ff.
   flow rates in, **7.**11 ff.
     for domestic water, **7.**12
     friction head loss, **7.**12
     load factor determination, **7.**12
   Hardy Cross network analysis method, **7.**15 ff.
   industrial water and steam requirements, **7.**20
   municipal water sources, **7.**17
   parallel and single piping, **7.**11
   pressure loss analysis, **7.**11 to **7.**16

Water-supply systems (*Cont.*):
　pump drawdown analysis, **7.1** to **7.9**
　　drawdown in gravity well, **7.4** to **7.9**
　　Dupuit formula in, **7.2**, **7.4**
　　recovery-curve calculation, **7.6** to **7.9**
　　wells in extended use, **7.6** to **7.9**
　selection of, **7.17** to **7.21**
　　fire hydrant requirements, **7.19**, **7.20**
　　flow rate computation, **7.17**
　　piping for, **7.17**
　　pressurizing methods, **7.19**
　　water supply sources in, **7.17**
　treatment methods, **7.21** to **7.24**
　　disinfection, **7.23**
　　filtration, **7.21**
　　softening, **7.22**
　　solvents, **7.23**
　typical municipal water sources, **7.17**
　water wells, **7.1** to **7.11**
　　air-lift pump for, **7.9** to **7.11**
Web reinforcement:
　of prestressed-concrete beam, **2.59**
　of reinforced-concrete beam, **2.9** ff.
Web stiffeners, **1.43**
Wedge of immersion, **6.4**

Weir:
　discharge over, **6.10** ff.
　variation in head on, **6.10**
Welded beams, **1.41** ff.
　design moment of, **1.43**
Welded connection, **1.43**, **1.120** ff.
　welded flexible, **1.151**
　welded moment, **1.154**
　welded seated, **1.154**
Welded plate girder design, **1.19** ff.
Westergaard construction, **4.32**
Williott displacement diagram, **1.65**
Wind drift reduction, **1.169** ff.
Wind stress analysis, **1.162** ff.
　cantilever method, **1.164**
　portal method, **1.162**
　slope-deflection method, **1.167**
Wood beam (see Timber beam)
Wood joist(s), **3.2**
Wood-plywood girder, **3.4** to **3.6**
Working-stress design, **2.35** to **2.39**

Yield moment, **1.122**
Yield-point stress, **1.122**

Zenith of observer, **5.12**, **5.13**